ETHICS FOR LIFE SCIENTISTS

Wageningen UR Frontis Series

VOLUME 5

Series editor:
R.J. Bogers
*Frontis – Wageningen International Nucleus for Strategic Expertise,
Wageningen University and Research Centre, Wageningen, The Netherlands*

Online version at http://www.wur.nl/frontis

ETHICS FOR LIFE SCIENTISTS

Edited by

MICHIEL KORTHALS

and

ROBERT J. BOGERS

A C.I.P. Catalogue record for this book is available from the Library of Congress.

ISBN 1-4020-3179-3 (PB)
ISBN 1-4020-3178-5 (HB)
ISBN 1-4020-3180-7 (e-book)

Published by Springer,
P.O. Box 17, 3300 AA Dordrecht, The Netherlands.

Sold and distributed in North, Central and South America
by Springer,
101 Philip Drive, Norwell, MA 02061, U.S.A.

In all other countries, sold and distributed
by Springer,
P.O. Box 322, 3300 AH Dordrecht, The Netherlands.

Printed on acid-free paper

Front cover illustration: Cyprian Koscielniak/NRC Handelsblad

All Rights Reserved
© 2004 Springer
No part of this work may be reproduced, stored in a retrieval system, or transmitted
in any form or by any means, electronic, mechanical, photocopying, microfilming,
recording or otherwise, without written permission from the Publisher, with the
exception of any material supplied specifically for the purpose of being entered
and executed on a computer system, for exclusive use by the purchaser of the work.

Printed in the Netherlands.

Contents

	Preface	ix

Introduction

1.	**Ethical challenges for the life sciences** *Michiel Korthals (The Netherlands)*	1

Researchers in organizations

2a.	**Moral complexity in organizations** *Ronald Jeurissen (The Netherlands)*	11
2b.	**Comments on Jeurissen: Organization and moral complexity** *Hugo Letiche (The Netherlands)*	21
3a.	**The social role of businesses and the role of the professional** *Johan Wempe (The Netherlands)*	27
3b.	**Comments on Wempe: Conditions for ethical business** *Henk Zandvoort (The Netherlands)*	41

Responsible authorship and communication

4a.	**The responsible conduct of research, including responsible authorship and publication practices** *Ruth Ellen Bulger (USA)*	55
4b.	**Comments on Bulger: The responsible conduct of research, including responsible authorship and publication practices** *Henk van den Belt (The Netherlands)*	63
5a.	**Professional ethics and scholarly communication** *Hub Zwart (The Netherlands)*	67
5b.	**Comments on Zwart: Professional ethics and scholarly communication** *Tjard de Cock Buning (The Netherlands)*	81
6a.	**Some recent challenges to openness and freedom in scientific publication** *David B. Resnik (USA)*	85
6b.	**Comments on Resnik: Some recent challenges to openness and freedom in scientific publication** *Tiny van Boekel (The Netherlands)*	101

Ethics of animal research

7a.	**Research ethics for animal biotechnology** *Paul B. Thompson (USA)*	105
7b.	**Comments on Thompson: Research ethics for animal biotechnology** *Mieke Boon (The Netherlands)*	121

Ethics for life scientists as a challenge for ethics

8a.	**How common morality relates to business and the professions** *Bernard Gert (USA)*	129
8b.	**Comments on Gert: Gert's common morality: old-fashioned or untimely?** *Jozef Keulartz (The Netherlands)*	141
9a.	**Research as a challenge for ethical reflection** *Marcus Düwell (The Netherlands)*	147
9b.	**Comments on Düwell: Research as a challenge for ethical reflection** *Akke van der Zijpp (The Netherlands)*	157

Scientists in society

10a.	**New public responsibilities for life scientists** *Michiel Korthals (The Netherlands)*	163
10b.	**Comments on Korthals: New public responsibilities for life scientists** *Jan H. Koeman (The Netherlands)*	171
11.	**Science, context and professional ethics** *Ruth Chadwick (UK)*	175
12a.	**Bioscientists as ethical decision-makers** *Matti Häyry (UK)*	183
12b.	**Comments on Häyry: Assessing bioscientific work from a moral point of view** *Robert Heeger (The Netherlands)*	191

New developments

13.	**The human genome: common resource but not common heritage** *David B. Resnik (USA)*	197

Conclusions

14. Towards ethically sound life sciences 213
Michiel Korthals (The Netherlands)

List of authors 219

Preface

In May 2003, a group of experts in the ethics of life sciences came together in Wageningen, The Netherlands, to discuss the most urgent ethical problems faced by these sciences and connected technologies. This volume is a reflection of the presentations and discussions during this Workshop. Tassos Michalopoulos (Applied Philosophy Group, Wageningen University) and, in particular, Wendelien Ordelman (Frontis, Wageningen University and Research Centre) have, by their continuous support and assistance, greatly contributed to the favourable context for these discussions. We thank the authors for their important contributions, and Wageningen University and Research Centre for providing the funds for this meeting. We also thank Paulien van Vredendaal (Library, Wageningen University and Research Centre), who played an important role in the lay-out process.

The editors,

Michiel Korthals
Robert J. Bogers

Wageningen, July 2004

INTRODUCTION

1

Introduction: Ethical challenges for the life sciences

Michiel Korthals[#]

The rise of ethics for the life sciences: Why?

Since World War II many feel the necessity of reflecting upon the values that are at stake when doing scientific and technological work in the field of the life sciences. In doing experiments with animals, plants or humans the welfare of these living beings can be hampered; in publishing articles issues of private and public concern can be harmed (patents!); in being a member of a research group issues of human rights (like discriminatory behaviour) can become prominent; individual and collective forms of responsibility because of controversial types of research can become urgent; funding organizations can confront scientists and engineers with new ethical issues; the public at large or, as is the case with sustainability, future generations can challenge existing ways of doing research and educating and teaching can confront scientists with new ethical issues.

In all these dimensions problems have emerged, and scientists and technologists are sometimes confronted with even undeserving negative criticisms. However, often something or someone was damaged by scientific and technological activities; in other cases there were at least lots of uncertainties about proper ways to act. So, in those cases, it is not necessary to draw a red line between the do's and don'ts, but rather to explore the uncertainties.

In journals like *Business & Professional Ethics Journal*, the *Journal of Business and Professional Ethics* and *Journal of Academic Ethics* exclusive attention is paid to these topics, and prominent societies of scientists and engineers have departments, study groups or networks devoted to discussing them. Several universities have established departments for conducting research and teaching in professional ethics, like the Illinois Institute of Technology, MIT or Carnegie Mellon University.

Science objective and ethics subjective?

However, can ethics be helpful in reflecting on these compelling problems with which the life sciences are confronted? It is often said that science is impersonal, objective and without values, and ethics is often seen as the counterpart of science, as something personal and subjective. If these characteristics are valid, ethics can only be helpful by giving some psychic relief to the individual scientist who feels now and then the pangs of his conscience. However, it is simply untrue that science is not committed to values, or that values are not functioning in science. In the process of doing scientific research lots of values are at stake: think about the values that govern the interaction with scientific colleagues, funding institutions or employers, and

---

[#] Applied Philosophy Group, Department of Social Sciences, Wageningen University, Hollandseweg 1, 6706 KN Wageningen, The Netherlands. E-mail: Michiel.Korthals@wur.nl

clients and users. Or think about the values that govern your answers in how far you can use animals or human beings in scientific experiments. Moreover, the products of science can have fundamental influence in the daily lives of many, like now is the case with medicine, genetics and agricultural sciences. Here the list is endless, and later on we will discuss the intricacies of these relationships between scientifically produced objects and societal aspects. Even the methods of science can have *prima facie* damaging effects on human and nonhuman beings, like in the case of certain ways of environmental accounting in cost–benefit analysis (Shrader-Frechette 1994).

Anyhow, all these examples show that between science and ethics there is a very complex, fundamental, problematic, ambivalent and sometimes even dilemmatic relationship that is absolute worthwhile to study (Keulartz et al. 2002).

Why do ethics?

Even if it the case that there is a complex relationship between ethics and science, for many it is still unclear what ethics (as a discipline) can contribute to these issues. Is ethics only dependent on personal insights? Can ethics be learned? First, it is very clear that lots of children learn ethical concepts (Van Haaften, Korthals and Wren 1997). With their birth, they do not have any ethical insight or intuition or competence in ethics, but more and more they acquire the competence to speak with their peers and relatives on rather simple ethical issues like dividing a birthday cake or sharing movie tickets even if there is not enough for everyone. So, during these years children learn new concepts enormously fast and get ethical skills in using them. Adults as well learn (in the sense of making qualitative new steps in their ethical performance) during maturing processes. Even organizations and other collectives can learn in an ethical way: take the way in the United States people have learned to eliminate more and more the most deeply controversial elements of the race-segregation system. Indeed, even nowadays there is a lot to be improved, but the improvements as compared with the situation of the nineteen fifties, when a large part of the majority was completely ostracized, are still encouraging for anyone who doubts that people can improve their interaction with others in an ethically sound way.

So, ethical education is possible and necessary and ethics can be learned, and ethical arguments can play a role in this learning process. Speaking about ethical arguments means saying that an ethical decision can be proven just like in science. More than in science, proof and argumentation on ethics are indeed dependent on personal and social contexts. Whereas in science repeating an experiment is a type of ultimate proof, in ethics, time elapsed is important and past events (actions) cannot be repeated in the same way: once you have lied to someone, you have done it, and only with the utmost effort (like apologizing) you can restore trust. Indeed in ethics we can have proof and certainty, but in a way different from science. Because learning ethics means acquiring attitudes, concepts, methods and heuristics, social contexts play a pivotal role. Social structures create many of our norms and values, and facilitate or prevent people's access to some norms and values and not to other ones. Indeed, ethics is socially and culturally dependent, but still something one can learn, by transgressing one's own context. An ethics course can help with this process.

What kind of ethics?

If ethics is indeed becoming more important than ever for the life sciences, the next issue is, what type of ethics do we need to tackle these problems? Given the innumerable different approaches in ethics, the large controversies and the lack of consensus on main issues, this looks like an unsolvable problem. One very fundamental question is, do we need an ethics that first formulates general principles and norms for everyone, than detail these for individual branches and professions, and finally spell out what this means for the professionals? Or do we need an ethics that is sensitive to the diversity of ethical problems, the wide variety of moral dilemmas and the most urgent ethical problems in the field, and then tries to elucidate the intuitions at hand, analyses the most common solutions and explicates what heuristics or guidelines could improve the ethical situation? In other words, if ethics can be helpful in tackling ethical problems of the life sciences, are we in need of a more deductive perspective or a more narrative and inductive perspective? Final answers to ethical problems in accordance with one of these broad perspectives need not differ much, although the reasoning and practices do as a matter of fact. For example, both perspectives can end up in underwriting a professional code for life scientists; however the function of this code can be seen differently. In this book this issue is dealt with from various angles and is subject to several considerations. Consensus can be found, but is sometimes difficult to reach and not always recommendable, because it is a barrier for improving the debate. Moreover, the life sciences are not a united field where the same types of problems pop up. In some case a more principled approach may really be fruitful, whereas in others a more inductive and narrative approach may be recommendable.

A last, but not least, fundamental issue concerns the value of non-scientists' opinions and preferences. Because of the large societal impact of the life scientists, it is as least worthwhile to consider consultations with the public at large, in particular on the issues already mentioned (Korthals forthcoming). The qualities of the process of ethical improvement could be enormously enhanced by, in one way or another, involving the concerns of consumers with respect to animal welfare, human health, the environment (like sustainability and biodiversity), fair trade and fair treatment of farmers in developing countries, food diversity and transparency and accountability of these sciences. In this book several ways of public consultations are vetted and assessed.

Topics of this book

In this book we will first discuss broader issues of ethics of the life sciences, which enable us later on to focus on the more specific issues. Therefore, we begin with two contributions on the ethical issues of working in organizations. A fruitful side effect of this start is that it gives a good insight into business ethics, a branch of applied ethics that until now is far ahead of ethics for life scientists. In the second part, ethics of activities directly connected with doing scientific research are discussed, like experimenting with animals and human beings, publishing, patenting, getting funds and selecting one's research theme. Thirdly, we discuss the topic of animal ethics, which in the life sciences in particular requires discussion on the use of animals in experiments. Fourthly, several authors present their view on the relationship between science and society, i.e., the societal impact of the life sciences, like genetic, agricultural and food research. Fifthly, we discuss the issue of the impact of this applied field of ethics for academic ethics in general. We will finish with some new problems with which the life sciences are confronted.

More in detail, the kick-off is Ronald Jeurissen's chapter on *Moral complexity in organizations*, which deals with problems of ethics management. The aim of his chapter is to bring the discussion on ethics management a few steps further, starting from the received and dominant view in ethics management at this moment, that there are basically two approaches to ethics management, namely a rules-based and a values-based approach. Jeurissen presents a contingency model of ethics management that enlarges this dominant view, by bridging a kind of divide that presently exists between ethics management and stakeholder management. Stakeholder management is generally seen as a tool of corporate social responsibility (CSR), and CSR is often seen as something distinct from ethics management. CSR is supposed to deal more with the external relations of the organization (the impact on stakeholders), whereas ethics management allegedly has more to do with internal relationships (employee conduct). Jeurissen believes that this is a fruitless distinction that actually blocks further progress in ethics management. He argues that instruments of ethics management can instead be ordered along a continuum of increased moral complexity. This continuum blurs the existing artificial boundaries between ethics management and CSR. The chapter, therefore, can be understood as an attempt to integrate ethics management and CSR, and to contribute to a more unified theory and practice in the field of applied ethics in organizational contexts.

In his comments Hugo Letiche concentrates on the concept of 'moral complexity' to describe the interactions between the four types of ethics in organizations that Jeurissen distinguishes. Complexity in the sense of complexity theory is at odds with any regime of fixed positions or 'rules', has to deny change to remain valid. This denial of complexity vitiates the 'rules' of lived authenticity. Complexity means emergence and self-organizing. Human institutions evolve and change from a dynamism that is outside human understanding or control. Emergence is a guarantee of change and activity – it ensures that there will be indeterminate situations to examine and possibilities for multiple courses of action.

The second chapter, written by Johan Wempe and Thomas Donaldson, addresses the ethical behaviour of organizations and their agents. They criticize a reaction to perceived unethical behaviour in organizations, called by the authors the 'compliance/market picture', which they claim is based on a too simplistic analysis of ethical misconduct in organizations. The authors argue that many cases of alleged organizational misconduct occur in situations characterized by a 'plurality of values', and they present a strategy for dealing with value conflicts arising from this. This strategy they call 'integrity'.

In his comments to Wempe and Donaldson, Henk Zandvoort first addresses the 'compliance/market picture' and what the authors say on that. He presents his own analysis of the nature of a very large class of alleged misconduct of business organizations, namely those cases having to do with environmental damage and all kinds of harm and nuisances which are commonly called negative externalities in the science of economics. Secondly, he discusses the authors' ideas regarding 'integrity' of business organizations.

In the next section communication with the public and with peers (written or in another form) is discussed. Ethical issues like what is an author, who is responsible for the publication and several other ones are discussed in depth.

Ruth Ellen Bulger identifies several areas in this field. They are as follows: acquisition, management, sharing and ownership of data; mentor/trainee relationships; responsible authorship and publication practices; peer review and the use of privileged information; collaborative science; research involving human volunteers; humane care and use of animals; research misconduct; and conflicts of interest and commitments.

The process of establishing norms for ethical conduct of research in these areas is far from complete. However, momentum is gained when the members of various disciplines form consensus groups such as the International Committee of Medical Journal Editors. In addition, the extensive availability of internet communication resources among scientists around the world is forcing re-evaluation of the traditional expensive method of publication with limited distribution of material published in the print media only to subscribers.

In his comments Henk van den Belt poses some sceptical questions. He argues that from the late 1970s on, the science journals and the mass media have been reporting an unending series of affairs involving fraud, deceit, plagiarism and other forms of 'misconduct', especially in the biomedical sciences. In fact this trend has also been the main factor behind the rise of research ethics, at first in the USA and later elsewhere. He is doubtful on the question in how far research ethics can offer useful solutions to do something about the problem.

Prof. Dr. Hub Zwart addresses the history of scholarly expressions, and in particular the second last stage where a new type of professional scientific activity (systematic observation, using sophisticated equipment, and notably experimentation) entailed the emergence of a new scientific genre: the research paper (as well as the scientific journal: a periodical compilation of research papers). In the last stage we experience now, commercialization and intellectual autonomy get a new impact. Moreover, the impact of the informational revolution, pluralism of ethical styles and the fairness or unfairness of citation practices requires something like the Vancouver Guidelines (a topic also discussed by Bulger). He concludes that nevertheless science is a powerful tool developed by man to educate and discipline him and that scientific training is basically training in self-control. Virtues involved in practicing a science, such as unprejudiced open-mindedness, patience, precision and reliability, are moral values.

The commentator, Tjard de Cock Buning, disagrees with Zwart on several points but the bone of contention is in how far ethics of life sciences should orient themselves to principles that are firmly established in academic ethics, like those of Rawls, or be sensitive to the special ethical issues and dilemmas that play a role in the field of professional ethics. De Cock Buning regards the principles of Integrity, Responsibility and Competence the main principles that can solve these special problems.

In *Some recent challenges to openness and freedom in scientific publications*, David B. Resnik argues that openness and freedom are two important ethical values that apply to scientific inquiry. This chapter discusses some recent problems arising from industry-sponsored research and the danger of bioterrorism, which threaten openness and freedom in scientific publication. Resnik also discusses some possible solutions to these problems of openness and freedom and the problem of data sharing.

In his comments, Tiny van Boekel wants to draw attention to the opportunities of involving industry in the research, resulting in a fruitful interaction between science and society in pre-competitive areas.

The next section is dedicated to animal-research ethics. Paul Thompson concentrates upon the experimentations with animals in the case of biotechnology. He sticks to the still very relevant, but often neglected, ethical question of Rollins: what are researchers' responsibilities with respect to the animals they use in research? From a pragmatic bioethical perspective, he argues that the key research-ethics questions demand a scientifically informed approach to animal welfare, which in turn demands an understanding of the interpenetration between ethics and animal-welfare science. He concludes with a discussion of how research-ethics committees can approach the evaluation of animal biotechnology in a more ethically satisfactory manner.

In her comment, Mieke Boon compares the functioning of DECs (DierExperimentenCommissies, animal-experiments committees) in The Netherlands, with the functioning of Animal Care and Use Committees (IACUC) in the United States as described by Paul Thompson. She argues that besides governmental rules, there are also local rules effective.

In the following section we discuss the role of the life sciences in society, and the responsibilities for both science and scientists that are connected with this role.

Michiel Korthals explains that in the next decades life scientists will become more than ever involved in public and private life of patients–consumers, because of the shifts towards individualized (instead of collective), preventive (instead of curative) and desire-driven (instead of technology-driven) health and food sciences. This means that the relationship between doing research, giving advice to industry, governments and patients–consumers, consulting the public and prescribing products, be they patents, products, information or advice, is getting blurred. Traditional concepts of individual, role, task and collective responsibility have to be revised. Korthals argues from a pragmatist point of view that the concept of public responsibility can help a lot in delineating new grey zones between doing research for governments or industry, giving advice, prescribing and selling products, and doing public consultation. The main issues are where new Chinese Walls (not Berlin Walls) have to be built between these activities to improve trust between life scientists and the public at large and how to organize research agendas and to decide upon research topics.

Jan H. Koeman moderates some of the main trends Korthals delineates; genomics is just one step in the development of the life sciences and it remains to be seen in how far it can mean something substantial for the individual consumer. However, he illustrates with new data the problems of the role-responsibility theory of scientists and pleas for a better, more reasonable crossing of the areas of doing research, giving advice and consulting the public. He concludes that integrity of the individual scientists is still the most important feature in crossing these areas.

Ruth Chadwick also addresses the purported current crisis of confidence in science. She explains this, first, in terms of not only often undesirable but also unpredictable effects of scientific developments; and, second, in terms of the commercialization of science. It has led to calls for greater accountability of scientists to the public. The question arises, however, as to whether professional ethics provides an appropriate framework within which to address the issues. In particular she argues that the debate in professional ethics concerning internal and external values needs to be explored in relation to science, with particular reference to the context in which scientific research is conducted.

Matti Häyry examines the idea that bioscientists should somehow participate in ethical decision-making in their field. He states that they have ample reason to do so, and although it is sometimes difficult to see how they could actually contribute to the assessment of their own work, he gives some hints for their task.

In his comment on the contribution of Matti Häyry, Robert Heeger focuses on Häyri's three criteria for moral acceptability, his caring about the definition of their key concepts, and his reference to ethical principles of a professional code. Heeger tries to show that such principles can play the hoped-for guiding role only if they are firmly rooted in the professionals' power of moral judgment.

In the next section we discuss the relationship of this type of applied ethics to ethics in general. Ethics in general can learn a lot from ethics of life sciences and the other way round, but in what sense? Bernard Gert explains his ideas on common morality, which should provide a framework on which all of the disputing parties can agree, making clear who is responsible for the disagreements, and what might be done to manage that disagreement. Not every moral problem will have a single best solution, that is, one that all equally informed impartial rational persons would prefer to every other solution. Common morality recognizes both the vulnerability and the fallibility of people. It includes (1) rules prohibiting acting, or attempting to act, in ways that cause, or significantly increase the probability of causing, any of the five harms that all rational persons want to avoid, and (2) ideals encouraging the prevention of any of these harms. It also includes (3) a two-step procedure for deciding when it is justified to violate a moral rule. This common morality can be applied to scientists as well as to accountants and lawyers who are employed by a company. They cannot remain silent if the company is doing something that is contrary to the standards that their profession is committed to and with regard to which they have special expertise.

In his comments, Keulartz brings some sceptical comments to the fore that Gert's neo-foundationalism triggers, i.e. his emphasis on one common morality. Keulartz argues that even if we assume that there is only one moral reality, it is highly unlikely that there is only one unique description of this universe possible. The rule-consequentialist orientation of Gert can do no more justice to moral judgments that are considered significant than all other consequentialist theories. The most important of these judgments are judgments about distributive justice and judgments about respect for autonomy.

Marcus Düwell argues that one central implication of the ethical issues of life-science research concerns the normative framework of free and informed consent that now is so dominant. Exclusively paying attention to autonomous decision-making leaves out central questions like what the respect is that we owe to each other, and what implications this has for new scientific developments in view. The range of moral rights has not to be restricted to negative rights, meaning that not only those measures are necessary that protect everyone against direct interference in the freedom of his acting. He suggests that there could also be positive rights, like that we owe to each other the support we need in order to be enabled to live a good life. The complexity of the impact that research has on our society and existence forces us to open the discussion about the normative framework of such an evaluation.

Commentator A.J. van der Zijpp also points to the difficulties with the liberal position, both in terms of the room for making decisions predetermined by the scientific community and in terms of ethical attitudes like virtue, perfectionism and care. She emphasizes that somehow the rights and obligations humans have towards others (humans and animals and nature for example) are to be to taken into account. This implies a holistic approach in science, which requires social interaction for decisions about trade-offs between unequal issues regarding planet, profit and people.

In the last section David B. Resnik raises some new issues connected with the recent analysis of the human genome and the ethical issues that emerge with the regulation of intellectual property and patenting. One of the influential arguments against patenting human DNA is that the human genome is the common heritage of mankind.

This essay argues that the human genome is not literally the common heritage of mankind, but that it is a common resource. Since the genome is a common resource, the patenting of DNA is morally acceptable, provided that we honour our moral duties related to the genome, which include duties of stewardship and justice. This essay also explores different aspects of the debate over benefit-sharing in genetics research.

In his conclusion, Michiel Korthals summarizes main issues, sketches some ways out and poses some future questions and challenges.

References

Keulartz, J., Korthals, M., Schermer, M., et al. (eds.), 2002. *Pragmatist ethics for a technological culture.* Kluwer Academic Publishers, Dordrecht. The International Library of Environmental, Agricultural and Food Ethics no. 3.

Korthals, M., forthcoming. *Before dinner: introduction into food ethics.* Kluwer.

Shrader-Frechette, K.S., 1994. *Ethics of scientific research.* Rowman & Littlefield, Lanham. Issues in Academic Ethics.

Van Haaften, W., Korthals, M. and Wren, T. (eds.), 1997. *Philosophy of development: reconstructing the foundations of human development and education.* Kluwer, Dordrecht.

RESEARCHERS
IN ORGANIZATIONS

2a

Moral complexity in organizations

Ronald Jeurissen[#]

Business ethics is applied ethics, says Velasquez (1992, p. 2). The application of ethics to organizational contexts can take two forms, which are both 'practical', but in different ways. The first type of application aims at analysing specific ethical problem types in organizations, in order to provide normative clarification and guidance. Examples of this are advertising ethics, the ethics of insider trading or the ethics of company restructuring. The knowledge base of this type of application is standard ethical theory, such as justice theory or virtue ethics. Insights from these general theories are applied to specific organizational contexts.

The second type of application of ethics to organizations aims at improving the decision-making processes, the procedures and structures in an organization, so that the operations of the organization are more geared towards ethical principles. The knowledge base in this case is organization theory and management science. Here we come across a whole range of ethics-based organizational instruments and tools, ranging from codes of conduct and ethical audits to all embracing methods, sometimes called 'strategies', for running an organization the ethical way. When we refer to the first type of applied ethics as 'organizational ethics', then the second type is best referred to as 'ethics management'. The basic question of ethics management is simply: "how do you manage ethics in organizations?"

This paper deals with problems of ethics management. The aim of the paper is to bring the discussion on ethics management one or perhaps two steps further, starting from the received and dominant view in ethics management at this moment, that there are basically two approaches to ethics management, namely a rules-based and a values-based approach. I will present a contingency model of ethics management that enlarges this dominant view, by bridging a sort of divide that presently exists between ethics management and stakeholder management. Stakeholder management is generally seen as a tool of corporate social responsibility (CSR), and CSR is often seen as something distinct from ethics management. CSR is supposed to deal more with the external relations of the organization (the impact on stakeholders), whereas ethics management allegedly has more to do with internal relationships (employee conduct). I believe that this is a fruitless distinction, which actually blocks further progress in ethics management. I will show that instruments of ethics management can instead be ordered along a continuum of increased moral complexity. This continuum blurs the existing artificial boundaries between ethics management and CSR. This paper, therefore, can be understood as an attempt to integrate ethics management and CSR, and to contribute to a more unified theory and practice in the field of applied ethics in organizational contexts.

[#] Institute for Responsible Business (EIBE), Nyenrode University, Straatweg 25, 3621 BG Breukelen, The Netherlands. E-mail: r.jeurissen@nyenrode.nl

Chapter 2a

The standard model of ethics management: rules and values

The standard model of ethics management starts from the distinction between rules and values. This distinction has been introduced by Lynn Sharp Paine, in a seminal Harvard Business Review Article called 'Managing Organisational Integrity' (Sharp Paine 1994). Sharp Paine (1994) herself applies the concepts of 'compliance' and 'integrity', but I find that a bit confusing, because the word 'integrity' already connotates the broader genus of ethical qualities and policies in organizations, and is now being used to name a specific approach to ethics management as well (a values-oriented approach to ethics policies). Opposing 'compliance' and 'integrity' this way can also lead to the false impression that compliance is actually not a genuine strategy of integrity, whereas I believe it surely is.

Roughly, there are at present three generally recognized approaches to ethics management:
- a rules-based approach, aimed at the implementation of specific ethical rules of conduct in the organization;
- a values-based approach, aimed a creating an ethical organization culture;
- a stakeholding approach, aimed at proactively integrating the rights and values of stakeholders into the policies and strategies of the organization.

The three strategies are presented here as 'ideal types', as theoretically discernible approaches to the management of ethics, each having their own very distinct characteristics. Probably no organization has ever adopted one of these strategies in its pure form. In actual practice, companies make all kinds of combinations.

Rules

The goal of a rules-based approach to ethics management is to promote norm-conform action in the organization. These norms can be imposed on the organization externally, as in the case of legal norms, but the norms can also originate from within the organization itself, for example in the form of a voluntary company code of conduct. Norm-conform action is promoted by increasing the control over organizational members. To ensure compliance with the rules, a punishment scheme is required as well.

There are obvious legal reasons for organizations to adopt a policy of legal compliance. The risk of litigation and liability has increased in the past decades, as lawmakers have legislated new civil and criminal offences, stepped up penalties and improved support for law enforcement. One can think of laws against bribery, insider trading, money laundering or the abuse of corporate opportunity by managers. Law is becoming increasingly complex, and it is only prudent to have the basic legal risks put together for employees in the form of a compliance code. In order to prevent legal prosecution, both for the organization and for individual employees, companies have developed ethics programmes to detect and prevent legal violations. Existing systems of internal control can be extended to give 'reasonable assurances' in the field of legal compliance.

Typical elements in a rules-based approach therefore are:
- communication of the standards to which employees must adhere;
- monitoring of employee behaviour;
- procedures to report deviant behaviour;
- disciplinary measures.

In most cases, either the legal department or internal audit is charged with the execution of the compliance programme.

Values

The law is an important ethical bottom line. But there is more to ethics management than legal compliance, for two reasons. Firstly there is a moral reason: the law sets only the lower limits of ethical conduct; it does not inspire to human excellence, to exemplary behaviour, or even to good practice. Secondly, there is an organizational reason: a programme of legal compliance does not address the root causes of misconduct in an organization. These are often organizational in nature. An organization sends many signals to its employees as to which behaviour is favoured by the organization and which is disapproved. These signals stem from both the formal and informal systems in the organization. To create an ethical culture, these systems must be aligned to support ethical behaviour. For example, if the formal ethics code tells people that honesty is highly valued in the organization, and high-level management routinely tells customers the truth about the organization's ability to meet their needs, employees receive a consistent message about the organization's commitment to honesty. The systems are aligned. On the other hand, if the same organization regularly deceives customers in order to land a sale, the organization is out of alignment. Its formal culture says one thing, while its informal culture says the opposite (Treviño and Nelson 1995, p. 197).

An organization that is serious about ethics must proactively develop an ethical culture. This is where organizational values come in. The roots of an organization's ethics are its guiding values, which make up the core of its culture. The values give the organization a framework of reference that gives guidance to the acting of officers, managers and employees. At the same time, the values contribute to the development of the identity of the organization. A values-based approach to ethics management aims at making the key values of the organization pervasive in all aspects of its behaviour, so that the organization sends one consistent ethical message in everything it does, internally and externally. A values-based approach to the management of integrity stresses the own responsibility of employees. The objective of this strategy is to enable and stimulate employees to make autonomous and well considered moral judgements. To this order, the organization formulates core values, which serve as guidelines – and not as rules – for the employees.

Sharp Paine (1994, p. 111) emphasizes that a values-based strategy poses higher demands to an organization than a rules approach. She says that a values approach is:

"(...) broader, deeper and more demanding than a legal compliance initiative. Broader in that it seeks to enable responsible conduct. Deeper in that it cuts to the ethos and operating systems of the organization and its members, their guiding values and patterns of thought and action. And more demanding in that it requires an active effort to define the responsibilities and aspirations that constitute and organization's ethical compass".

Comparison

Karssing (2001, p. 27-35) points out that a values approach is often more functional than a rules approach, because ethical problem situations in organizations are often too complex to be captured in uniform rules. In particular, he points to the following limitations of organizational rules:

- Rules and prescripts are necessarily of a general nature, which leaves little room for variance and which can lead to rigid policies. There are always more situations than rules.
- The meaning of rules is not always clear, so that it is not always obvious whether a specific situation does fall under a certain rule or not. Remember the famous case of this lady who went out to walk her pet tiger, and came across a shield in front of a park, saying "no dogs allowed".
- In a dynamic world, rules often lag behind social and technological developments, which means that today's actions are guided by yesterday's rules.

Formalising Karssing's comments, we can say that the two approaches are distinguished by a difference in the degree of moral complexity. Moral problems in organizations are sometimes too complex to be handled by rules in a proper way. When the management of ethics calls for more generalized normative frameworks, because of the complexity of the action situation, a values-based strategy is more suited then a rules-based strategy. This implies that rules are not always a second-best solution to integrity management. Where action situations are unequivocal, both in a descriptive and in a prescriptive sense, an approach based on rules can often be functional and sufficient. A values-based strategy can even be disfunctional in such situations, because it lacks the necessary prescriptive precision. It is not enough, for example, to steer the behaviour of investment bankers regarding their private dealings in stocks by the general values of 'respect' and 'integrity'. Greater precision is required here, both in the description of specific situations and in the prescription of specific actions.

Broadening ethics management

Several authors have tried to broaden the conceptualization of ethics management, beyond the binary scheme of rules versus values. At this point in time, the development of theory is still in an exploratory phase, but it is remarkable that many initiatives point to a single direction. Many researchers in this field believe that the theory of ethics management should pay more attention to the relationship between the organization and its external environment, through stakeholder management and stakeholder dialogue.

In their comparative research of approaches to ethics management, Trevino, Weaver, Gibson and Toffler identify an 'external stakeholder' strategy, next to the existing strategies of rules and values. They say that: "(...) many companies hope to maintain or improve their public image and relationships with external stakeholders by adopting an ethics/compliance program. Therefore, we identified an orientation toward satisfying external stakeholders (customers, the community, suppliers) as a third approach (...)" (Treviño et al. 1999, p. 136). Unfortunately, this is all they have to say about the third strategy, and their theoretical account of it is therefore rather thin. The third strategy of Trevino et al. (1999) seems to be noting more than a new goal for the already existing strategies, namely the goal to satisfy stakeholders. A new goal is not the same as a new method, however.

Hummels and Karssing (2000, p. 201) also identify a third approach to ethics management, which they refer to as a 'facilitating' strategy. Here as well, the relationship with external stakeholders is central. "Facilitation is a strategy whereby the management not so much directs, motivates and steers, as well as listens to the stakeholders of the organization (...). The facilitation strategy does not ask 'What are the rules?', or 'Which actions are in line with our values?', but it asks 'What is mandatory in this particular context?' A facilitation strategy emphasizes the receptiveness to the viewpoints, interests and values of others ('responsiveness'). Central are *dialogue* and *learning* in and by the organization. The facilitation strategy is predominantly a process-oriented strategy. Central is neither how right the rules and regulations are, nor how right the leadership of the organization is; central is the uncertainty about the ethical course of the organization".

A stakeholding strategy bears witness to the fact that organizations do not stand alone in this world; they are surrounded by parties whose interests are at stake in what they do, and who can influence their performance. This simple fact has paramount implications for the ethics of an organization, and the way ethics is organized. Even an organization that is strongly values-driven cannot 'invent' its ethics on its own. How business affects its stakeholders and how it should best consider their rights and interests, is something that an organization cannot decide by itself. The stakeholders have an important say in this. What the business world has learned from incidents like Shell and the Brent Spar, Heineken in Burma and the working conditions at Nike's subcontractors in the Third World, is that even strongly values-driven companies can act in ways that are morally unacceptable to relevant stakeholder groups.

A stakeholding strategy is *responsibility-driven*. 'Responsibility' entails the word 'response'. Openness and preparedness to be held answerable and accountable by stakeholders are the keywords of the stakeholding approach. Trust and credibility in the eyes of the stakeholders are the rewards.

A historical benchmark of stakeholding is the Body Shop's social audit scheme. The Body Shop considers the engagement of stakeholder representatives in dialogue to be one of the most important and sensitive elements in its ethics programme. Stakeholders are consulted with a view to identifying the issues that are critical to the Body Shop's social and environmental performance and assessing the organization's performance against stakeholder-specific needs.

To a certain extent, the integrity strategy already bears witness to stakeholder values. By identifying 'integrity', 'honesty' or 'respect' as basic values, organizations implicitly express a fundamental sense of obligation towards stakeholders. But the question remains unanswered what specific commitments towards stakeholders follow from these values. This the values themselves do no tell. In a dialogue, the organization and its stakeholders together can identify the social role and responsibilities of the organization.

Theoretical integration: increased moral complexity

How do the three approaches of ethics management relate to one another? The foregoing overview of the three types suggests a contingency relationship, which to a certain extent can be seen as an evolutionary relationship. The evolution is driven by increased openness, increased trust and increased communication.

For many managers who are confronted with ethical problems in their organization for the first time, the instauration of a rule seems a probate means to tackle the problem. Managers have three obvious reasons to try to solve an ethical problem with recourse to a rule. Firstly, the rule can be communicated unequivocally, which holds the promise of clear guidance and control. Secondly, a rule can be proclaimed overnight, which holds the promise of speed. Thirdly, the manager who has put a rule in place can show to his superiors that he enacted his responsibility and has done something, with relatively little energy. This holds the promis of results. No wonder that a rules-based approach is at this time the most popular instrument of ethics management in the United States.

It requires a good deal of organizational learning for a company to step up from rules to values. Lynn Share Paine (1994, p. 112) points out that a successful values strategy requires an active effort to define the values and aspirations that constitute an organization's ethical compass. The guiding values must be clearly communicated. Company leaders must be personally committed, credible and willing to take action on the values they espouse. The espoused values must be integrated into the normal channels of management decision-making. The company's systems and structures must reinforce its values. And all partners in the organization must have the decision-making skills, knowledge and competencies needed to make ethically sound decisions on a day-to-day basis.

The step towards stakeholding involves a second cycle of organization learning. The organization must learn to look at itself through the eyes of its stakeholders. The scope of communication must be extended from the organization to the entire forum of its stakeholders. This involves an act of 'letting go' and surrendering, which runs counter to the prevailing ethos among managers that more control is always better. Stakeholders cannot be controlled. But stakeholder relations can be managed. Through dialogue processes organizations and their stakeholders can build up relationships of mutual trust, based on openness, accountability and interdependence.

A downside of the evolutionary view of the three strategies is that it seems to imply that stakeholding is always 'better' than a values-based strategy, which on its turn is supposed to be better than a rules-based approach. I believe this is a very erroneous way of looking at ethics management. In fact, there is no general preference for any of the three approaches, from a practical perspective. In actual practice, each of the three approaches can be an adequate solution to specific types of ethical problems, depending of situational characteristics. In order to capture these situational characteristics, I propose an integrated model of ethics management that is based on the increased moral complexity of situations that present themselves to organizations as ethical problems. The moral complexity of the situation is reflected in the complexity of the ethics management that tries to answer it.

The moral complexity of the situation involves two dimensions: the action context and the normative context. The *action context* becomes more complex, the more the ethical problem extends over different and more heterogeneous forms of action. The possibilities of shaping an ethics policy based on a closed set of rules are thus reduced. There are always more situations than rules. To answer this increase in action complexity, the instruments of ethics management need to become more generalized and more open. The steering becomes more dynamizing and motivational and less aimed at specific descriptions and precise control. An increase of action complexity calls for a generalization of normative frameworks, that is to say a shift from specific rules to broader values.

The *normative context* becomes more complex when ethics policies have to take different, and more diverging, normative frameworks into account. This way the applicability of a rules-based approach to ethics is reduced as well, and the need for a more generalized approach emerges. An increase of normative complexity is related to the involvement of more, and more heterogeneous, stakeholder groups, who represent diverging and even competing normative frameworks and world views. Hence, the increase of normative complexity is related to the multiculturality or transculturality of the environment in which the organization tries to answer ethical problems. Here, the generalization will be found in particular through legitimizing procedures that are applicable in a great number of different situations, notably the dialogue with stakeholders and the intercultural dialogue.

When we label the action complexity and normative complexity as 'low' and 'high' respectively (for reasons of theoretical simplicity), we obtain an analytical distinction of four types of ethical problems and four corresponding types of ethics management: a rules-based approach, a values-based approach, a stakeholder-dialogue approach and finally a social-dialogue approach (see Table 1).

Table 1. Ethics management and increased moral complexity

		action complexity	
		low	high
normative complexity	low	rules (1)	values (2)
	high	stakeholder dialogue (3)	social dialogue (4)

(1) Rules

A rules-based approach to ethics management is characterized by a combination of low action complexity and low normative complexity. The ethical problem pertains to a clearly demarcated set of actions, and all parties involved base themselves on the same set of moral criteria to assess the situation. This is the case, for example, when financial-asset managers of a bank have to ask themselves how they should deal with private transactions in stocks. The adequate regulatory answer to these types of problems is a set of rules. The fact that the rules involved can be of quite some technical complexity does not alter the low complexity as it is understood here. The key is that a clear complete description of the moral situation and the normative guidance is possible. Once the rules are understood, they are clear and it is possible to make such rules in principle.

Typical examples of a rules-based approach to ethics management can be found in the Business Conduct Guidelines of IBM (IBM Nederland N.V. 1998). This document entails twenty-three pages of detailed behavioural requirements and prohibitions for IBM personnel on a whole range of topics, such as the regulation on supplying to IBM:

"Generally, you may not be a supplier to IBM, represent a supplier to IBM, work for a supplier to IBM, or be a member of its board of directors while you are an employee of IBM. In addition, you may not accept money or benefits of any kind for any advice or services you may provide to a supplier in connection with its business with IBM. Also, you may not work on any products or services offered by a supplier to IBM."

This example does not mean to defend that IBM chose the appropriate ethics strategy for guiding their employees in the matter of supplying to their own company. Ethics officers should carefully consider what ethical topics in their organization need regulation in the first place, and they should establish that situations are both descriptively and normatively unequivocal, before installing a rules-based policy. A drawback of IBM's rules approach to the supplying issue could be that employees become trapped in the regulatory format, in a specific situations. The word "generally" suggests, however, that exceptions to the rules may be possible, in some situations.

(2) Values
When the action complexity is high and the normative complexity is low, we are dealing with situations where employees are expected to come up with creative new solutions to new ethical problems, however within a fixed normative framework. The normative framework should be open for transfers to new situations and therefore it should be of a general nature. An open formulation of organizational values can function as guidelines to orientate the acting of employees. The specific implementation and operationalization of the values are left to the own responsibility of organization members. A good example of this type of ethics management is the famous credo of Johnson & Johnson Pharmaceuticals (2004), where we can read:

> "We are responsible to the communities in which we live and work and to the world community as well. We must be good citizens – support good works and charities and bear our fair share of taxes. We must encourage civic improvements and better health and education. We must maintain in good order the property we are privileged to use protecting the environment and natural resources" (Johnson & Johnson 2004).

(3) Stakeholder dialogue
Low action complexity, in combination with high normative complexity, occurs in particular when organizations are faced with strong criticism on a specific 'issue', for example from the side of a non-governmental organization. It is clear to everyone what the issue is about, what the behavioural alternatives are, but there is no consensus about the legitimacy of the alternatives, because the normative frameworks are different. In a dialogue, the organization and its stakeholders can try to understand their normative viewpoints, to change perspectives and perhaps to change their own views. One step further, the organization can even enter the dialogue with stakeholders in the initial phase of policy development. An example of this is the extensive stakeholder-dialogue process organized on behalf on Shell, about the disposal of the Brent Spar. In a series of consultations in 1997 in London, Copenhagen, Rotterdam and Hamburg, Shell discussed alternatives for the controversial deep-sea disposal of the Brent Spar with a broad spectrum of stakeholders. The consultations contributed to the selection of a new disposal option of the Brent Spar, which was to be the integration of sections of its hull into a quay alongside the southwestern Norwegian town of Stavanger.

Stakeholder dialogue is a procedural generalization, to cope with high forms of normative complexity. It is a method which can be applied time and again, to reach legitimate solutions in normative conflicts with stakeholders.

(4) Social dialogue
Social dialogue can give answers to ethical problems where both the action complexity and the normative complexity are high. This involves situations where people have great difficulty to reach agreement, and even understanding, because the definitions of both the objective and the normative world diverge. Such morally ultra-complex problems occur to organizations in particular with regard to problems where there are great ideological (normative) differences in the society, coupled to a high amount of scientific controversy. Examples of this are the debates about nuclear energy, genetic modification or human-rights issues in an intercultural context. The disagreement around these issues leads to a polarization among the stakeholders, which strongly complicates the stakeholder dialogue ("damn if you do, damn if you

don't"). Individual stakeholders cannot provide legitimacy to the organization, because of the socially controversial nature of the problem. Therefore, the discussion has to be brought to the higher level of a societal dialogue. The organization and its stakeholders then become jointly responsible for an adequate handling of the problem.

The problem of moral ultra-complexity poses itself in particular for multinational organizations that operate in different cultures. The social dialogue then becomes an *intercultural dialogue*. Within each separate cultural realm, the moral complexity may be small, but it increases dramatically at the global level, because of the need for the multinational organization to account for its doings before a global forum of stakeholders. Which cultural framework should be taken as the point of reference then? The organization is torn between ethical relativism and universalism (Donaldson 1996).

Conclusion

The contingency view on ethics management outlined here implies that none of the four approaches to ethics management is self-sufficient. They augment and mutually support each other, each contributing to the ethical performance of the organization in a way that the others do not. No larger organization can afford not to have a clear compliance strategy in place, in order to cope with the ever-increasing burden of legal risks. In addition, a growing number of companies see the importance of an ethical corporate culture, as a sustaining foundation of legal compliance and as a source of ethical excellence 'on top of' the law. Collins and Porras have shown that companies that have a strong corporate culture and a strong sense of corporate identity are often the global winners in their industry, in terms of strategic foresight, flexibility and profitability. Clear ethical values are part and parcel of the culture of these winning companies (Collins and Porras 1995). Finally, there are companies that understand that they will prosper most when they are in balance with the world in which they live and operate. Knowing what the firm's social responsibilities are, through a continuous dialogue with stakeholders and in the confrontation between diverging frameworks of meaning, is the best way to ensure the legitimacy and social approbation of the organization.

References

Collins, J.C. and Porras, J.I., 1995. *Built to last: successful habits of visionary companies*. HarperBusiness, Century, London.
Donaldson, T., 1996. Values in tension: ethics away from home. *Harvard Business Review*, 74 (5), 48-62.
Hummels, H. and Karssing, E.D., 2000. Ethiek organiseren. *In:* Jeurissen, R. J. M. ed. *Bedrijfsethiek: een goede zaak*. Van Gorcum, Assen, 196-226.
IBM Nederland N.V., 1998. *Business conduct guidelines*. Available: [http://www-1.ibm.com/partnerworld/pwhome.nsf/weblook/guide_index.html].
Johnson & Johnson, 2004. *Our credo*. Available: [http://www.jnj.com/our_company/our_credo/].
Karssing, E.D. (ed.) 2001. *Morele competentie in organisaties*. Van Gorcum, Assen.
Sharp Paine, L., 1994. Managing for organizational integrity. *Harvard Business Review* (March 1-April), 106-117.
Treviño, L.K. and Nelson, K.A., 1995. *Managing business ethics: straight talk about how to do It right*. Wiley, New York.

Treviño, L.K., Weaver, G.R., Gibson, D.G., et al., 1999. Managing ethics and legal compliance: what works and what hurts. *California Management Review,* 41 (2), 131-151.
Velasquez, M., 1992. *Business ethics: concepts and cases.* 3rd edn. Prentice Hall, Englewood Cliffs.

2b

Comments on Jeurissen: Organization and moral complexity

Hugo Letiche[#]

Introduction

Jeurissen (in this volume) has attempted in "Moral complexity in organizations" to map the applications of ethics to organizations. What results is a four-part classification of organizational ethics, which I have (below) summarized in a Greimasian semiotic square (Greimas 2002). While some of the relations that are revealed are worked out in Jeurissen's original text, some are not. The 'move' of putting the analysis in this form is my responsibility. Jeurissen has introduced the concept of 'moral complexity' to describe the interactions between the four types of ethics in organizations which he has identified. I wish to develop this concept further, adding a complexity-theory interpretation of it. Jeurissen has assumed a common sense understanding of 'complexity' and 'organization'; I will complexify this.

Classification of ethics in organization

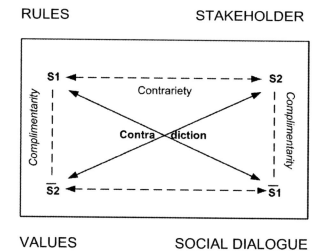

[#] University for Humanist Studies, P.O. Box 797, 3500 AT Utrecht, The Netherlands. E-mail: h.letiche@uvh.nl

M.J.J.A.A. Korthals and R.J. Bogers (eds.), *Ethics for Life Scientists*, 21-25.
© 2004 *Springer. Printed in the Netherlands.*

For Jeurissen the standard form of ethics in organizations is that of 'rules'. The distinction is based upon the difference between 'rules' and 'values'. As rendered above both 'rules' and 'values' share the assumption that basic ethical norms can relatively unproblematically be identified. In a regime of 'rules' it is assumed that a clear action agenda can be defined; and in a regime of 'values' it is assumed that actual action points are increasingly difficult to determine. The difference is in the difficulty of operationalizing the ethics, not in the assumed clarity of the ethical choices. If one can move easily from principle to action, 'rules' suffice. If it is difficult to move from ethical consensus to action, then 'values' are more appropriate. It is worth noting that a pragmatic approach to ethics holds sway here. One could argue that one should always choose for 'values', if possible, above 'rules'; for instance, because 'values' involve more consciousness of action and self-reflexivity. But Jeurissen makes no such choice. His criterion of selection is organizational performance - if 'rules' deliver fail-proof results, then they are preferred. If the uncertainties of operationalization demand adaptation, flexibility and tailored measures, organizational ethics can best be based on 'values'.

Jeurissen notes that the 'rules'/'values' dichotomy leaves out some crucial eventualities. What do you do if the 'rules' are as such, potentially clear enough; but you are confronted with a multitude of contradicting possible 'rules'? Stakeholder dialogue is proposed as the answer. As the semiotic square reveals, this may (or may not) work as a pragmatic solution, but it most certainly sacrifices the sense of 'values'. Ethics is reduced to a negotiated result that the stakeholders can live with. But this can easily be, in terms of values, a 'lowest common denominator'.

Yet more ambiguous, ethics and action possibilities can both be problematic – social dialogue is introduced as the response. Here, the logic of 'rules' does not apply, and 'value' clarity is 'missing'. Value consensus has to be found or defined. Social dialogue looks like stakeholder analysis, but in reality it answers to a far more philosophical agenda. Social dialogue presupposes that the 'truth' of ethics is unclear. While stakeholder analysis offers a political solution to organizational ethics, social dialogue opts for a 'coherentist' solution. (Bonjour 2000; Lehrer 1990; 2000) 'Coherentism' defines 'truth' in terms of what the members of the community can accept as true, while in principle respecting what they already recognize to be true. Social dialogue is focused on 'values', but it acknowledges the living community and the processes of its activity. Social dialogue is minimalist qua 'rules' – the only prerequisite is openness to examine, consider and evaluate the situation. Obviously, in social dialogue there are strong echoes of Habermas, and in its organizational ethics of Deetz (Habermas 1984; Habermas and McCarthy 1985; Deetz 1992).

Jeurissen proposes a contingency approach to the four ethical strategies. But a contingency solution is anything but ethically neutral. It involves, de facto, the reign of 'performativity'. But why should organizational or economic efficiency and effectiveness take precedence over more socially ethical criteria, such as justice or human flourishing?

Complexity theory and organizational ethics

The progression from rules to values, onto stakeholder dialogue, and to social dialogue, follows a logic of complexification. Jeurissen's definition of 'complexification' is framed in terms of complicatedness. In complexity theory the distinction is made between 'complicated' – that is, constituted of many elements; and 'complex' – that is, characterized by increasing levels of evolutionary development. Complexity theory is an evolutionary theory – it addresses the growing evolutionary or historical density of human existence. It speaks to the growing involvedness of the interrelated networks of living processes. Complexity addresses the changes, for instance from atom to molecule, compound, amino acid, one-cell life, higher levels of life, conscious life, society etcetera. At each level (new) properties emerge that could not be predicted on the basis of the properties at the lower preceding level. Complexity theory studies emergence, or the heart of the developmental progression from the most primitive life form to much more multifaceted organization, and onto contemporary society. As Edgar Morin, the sociologist of complexity theory, has argued, organization emerges out of disorder (Morin 1981). The principle of organization works in direct opposition to entropy – organization constantly absorbs more and more energy, in order to become ever more dense, dynamic and productive. Organization has become, in the last century, ever more complex – creating more levels of administration, more layers of meaning, increased technological sophistication and greater than before cultural differentiation. More and more 'difference' has been drawn into the structure(s) of organization – that is, into companies, multinationals, nations, cultures and globalization. The destruction of 'otherness', inherent to the making of organization, costs energy. Humanity has invested much of its will and passion, innovative capacity and dynamism, into creating ever more complex organization. But if this is 'good' is an altogether other matter. Sociability has become ever more complex – but as everyone from Foucault to Elias and from Serres to Morin has made clear, this is at a price. In nature, disorganization – emptiness, the undefined, the chaotic – is the rule, and order is the exception. Order, rationality and structure are always temporary – all advanced order dies; beginning with the humans themselves and including all that they create. Thus, the ontology of organization is itself paradoxical. Ever more powerful and performative levels of structure are built – the one above the other – and human energy is invested at a tremendous rate in ordering, structuring and the making of organization. And all of this perishes, and is overcome by emergent change and evolution. Organization is not something self-evident – it is a vast powerful human creation, which humanity has to invest energy to create and maintain. The 'human productive systems' that are necessary to maintain organization are mainly not experienced – it is only when there is a power failure that electricity is 'experienced'. Organization disappears into the 'black box' of the assumed and pre-reflective. The products and effects of organization are often more visible than their source in organizing.

Given that organization is ontologically, radically manmade, how should we think about its moral complexity? Organization consumes human energy – it sustains itself by taking much of what mankind is capable of, and internalizing that creativity, passion and ability, into itself. Human emergence – that is human resourcefulness, potential and vigour – is to a large degree absorbed by organization. In complexity theory, organization is a creation of evolution; but in management studies, organization determines evolution. 'Management studies' pretends that organizations can, via strategy, mission and tactics, determine events. But evolution has produced, and is producing, the ever more complex networks of interaction that provide the higher levels of control and order. Organization evolves. 'Business studies' portrays evolution as linear, rational and under control. But in reality, organizational and economic change are emergent. The technological and economic boom of the 1990s was not planned or predicted, nor was its end intended or foreseen. The technologies that drive economic growth, and the inefficiencies that create recessions, are all emergent. Complexity theory is the theory of emergence. Complexity theory acknowledges that change is non-linear, often unexpected, and frequently has unforeseen effects. Thus for a complexity-informed theory of organizational ethics, organizational ethics has to be social dialogue or it is nothing. Only social dialogue can acknowledge the relativity and difference inherent in the complex networks of our society. And only social dialogue can deal with the ethical problem of organization. Organization is not 'good' – it is complex, multifaceted and thoroughly saturated by 'will' and power. We all know that its possibilities are enormous, but so is its repressive potential. If organizational ethics assumes organization, and then fills in the ethics, it has reversed cause and effect. One needs to start with the human experiential level, and then move onwards to organization, as a complex human creation. The one thing we know for sure is that emergence will occur. Organization changes. Our need to assess the significance and import to us – as individuals and as communities – of that change, remains. Only social dialogue has the ability to acknowledge complexity, and what it signifies for organization.

Conclusion – Returning to the semiotic square

Social dialogue has the potential to be coherent in its answer to emergence. Any regime of fixed positions or 'rules' has to deny change to remain valid. This denial of complexity vitiates the 'rules' of lived authenticity. The 'rules' remain imposed absolutes – 'laws' that deny living organization and the principles of human evolution. 'Values' leave some crucial space for social processes, but are far too anthropomorphic. Again the radicalness of emergence is denied – human identity, manmade institutions and organization are all (more or less) self-organizing. That is, they evolve and change from a dynamism that is outside human understanding or control. Stakeholder determination reduces complexity to a political process of negotiation between pre-existing forces, positions and interests. But what complexity is all about is that emergence makes changes in the physical, natural, cultural and socio-economic order. An underlying acceptance of the fundamental power of the human life-force to create but also to alter organization, to make human projects possible but also to frustrate them, to define pockets of predictability but also to embrace unexpected change, is a prerequisite for a complexity-based ethics. Emergence is a guarantee of change and activity – it ensures that there will be indeterminate situations to examine and possibilities for multiple courses of action. Emergence defines the necessity of ethics – but a complexity-bound ethics.

References

Bonjour, L., 2000. The elements of coherentism. *In:* Bernecker, S. and Dretske, F.I. eds. *Knowledge: readings in contemporary epistemology.* Oxford University Press, Oxford, 128-148.
Deetz, S.A., 1992. *Democracy in an age of corporate colonization.* State University of New York Press, Albany.
Greimas, A.J., 2002. *Sémantique structurale: recherche de méthode.* PUF, Paris.
Habermas, J., 1984. *The theory of communicative action. Vol. 1. Reason and the rationalization of society.* Heinemann, London.
Habermas, J. and McCarthy, T., 1985. *The theory of communicative action. Vol. 2. Lifeword and system: a critique of functionalist reason.* Beacon Press, Boston.

Lehrer, K., 1990. *Theory of knowledge*. Routledge, London.
Lehrer, K., 2000. The coherence theory of knowledge. *In:* Bernecker, S. and Dretske, F.I. eds. *Knowledge: readings in contemporary epistemology*. Oxford University Press, Oxford, 149-165.
Morin, E., 1981. *La Methode. Tome 1. La nature de la nature*. Points Essais, Paris.

3a

The social role of businesses and the role of the professional

Johan Wempe[#]

Introduction

For years, the business sector was held up as an example for government organizations. Businesses work much more efficiently and are much more effective. This trend resulted in a profusion of privatization operations, as well as pressure on governments to deregulate. Over the last few years, however, the image of trade and industry has suffered serious damage. This was already an issue during the mid-'90s, with Greenpeace's criticism of the sinking of the Brent Spar, and snowballed as a result of the affairs concerning Enron, Ahold and Parmalat as well as the discussion of top salaries. All these debates have seriously lowered the level of esteem for the business world. A new aspect is the role of independent supervisory organizations that publicly correct those companies. Also striking is the changed role of what was traditionally designated as the social centerfield. Social organizations openly call businesses to account for their social impact. Other social organizations focus, on the other hand, on seeking partnerships with companies in order to realize social goals together. We are seeing considerable shifts, whereby the changing position of businesses plays a central role. Twenty years ago, it was unthinkable that anyone would speak of the human-rights policy of a company. That was considered an issue for governments. What's going on?

First of all, we shall discuss three examples of social debates in which businesses play a crucial role. There is talk of a structural problem for businesses that lies at the basis and, in fact, for society as a whole.

Top salaries

In various Western countries, a discussion has broken loose on the salaries of the top management echelon. In The Netherlands, the board of the Dutch Heart Foundation reconsidered the salary that was paid to the director of the foundation, the cardiologist V. Manger Cats, when it became evident that volunteers refused to collect donations for the Heart Foundation. According to the estimate of the Dutch Heart Foundation, the commotion concerning the salary of their director was responsible for a 12.5 % decrease in the annual house-to-house collection for the Heart Foundation (NRC Handelsblad, 21 May 2004). The ING Bank even received criticism from the Prime Minister for what he considered an exorbitant salary increase for the bank's administrators. And the recently appointed top man Anders Moberg of Ahold buckled under, and adapted his terms of employment when threatened with a consumer boycott. In Great Britain, the shareholders blew the whistle on chemical giant GlaxoSmithKline when it wanted to grant one of the top people an extravagant salary.

[#] Rotterdam School of Management, Erasmus Universiteit Rotterdam, P.O. Box 1738, 3000 DR Rotterdam, The Netherlands. E-mail: j.wempe@fbk.eur.nl

In the USA as well, the salaries and bonuses of various top entrepreneurs were the subject of a social debate.

The critics often point out the sharp contrast with the meagre results and the considerable price drops that the businesses have shown in recent times. Employees must tighten their belts. In a number of cases, this has even led to radical reorganizations and even dismissal of employees. Top managers who go home with a substantial salary increase give the wrong example, or enrich themselves at the expense of the employees, it is reasoned. It is not only about the level of the salaries, but what is also called 'a win-win situation'. When appointed, a top manager not only insists on a top salary, he also demands a golden handshake if it goes wrong. Even if he does not achieve anything, he still leaves with a pile of money.

In defence, it is maintained that the Board of Commissioners determine the salaries. In order to do so, they often set up a remuneration committee, which looks at developments among similar international companies. The world is the playing field of multinationals. In order to attract managers who can run an international company, you must offer attractive salaries in the international playing field. If, for a company such as Ahold, which is on the edge of the abyss after the fraud affair, you want to attract a new CEO who can save the company from ruin, you will have to dig deep into your pockets. Another argument in this regard is the high casualty risk. Today, if a top entrepreneur's performance drops, he will get sacked just as easily. Then just let him (or her) try to reach a similar level again. We must also point out that what we really want is to dispense with the option arrangements, whereby the administrator's income depends on the rise in prices. These arrangements are considered an important cause of the creative bookkeeping of companies such as Enron, WorldCom and Ahold.

What is an acceptable salary for a top manager? For the general public, any amount with five or more zeroes is a top salary. The general public is not able to follow the considerations of the Board of Commissioners of Ahold or the administration of the Heart Foundation. Ahold has weighed a number of possibilities and is happy to be able to draw in a top manager who is able to save the company that is at the edge of the abyss. The administration of the Heart Foundation is happy that they were able to attract the cardiologist who can speak on equal terms with the researchers who receive money from the foundation. That cardiologist has passed over a royal income as a medical specialist in order to accept the position of director.

However, the entrepreneur is increasingly becoming a public personality. The collector must be motivated to collect door-to-door for the Heart Foundation. The consumer must be prepared to pay the somewhat higher price of Albert Hein products. With Albert Hein stamps, the broad public saves for Ahold shares. People place a direct relationship between their own situation and the conduct of the top manager. They want to be able to trust him. The public wants the leaders of organizations who are responsible for business decisions that bring about major social consequences to think from a social perspective as well. In this respect, entrepreneurs are beginning to resemble ministers. They are expected to occupy their position purely in the public's interest. A personal advantage damages that ideal. A too high salary and other personal advantages are quickly seen as misuse of the position of the office.

For businesses, this discussion brings up a difficult problem. Negotiations are conducted with those involved on the terms of employment. The cost–benefit analysis is made behind the closed doors of the Board of Commissioners. Only the Board of Commissioners has any insight into the 'labour market for top managers'. At the same time, account must be taken of public opinion. The transparency that is required within the framework of Governance results in everyone looking over the shoulders of the Board of Commissioners. Can and may the Board of Commissioners allow themselves to be led by the perception of the broad public?

Honest coffee

Pressure groups call upon the major coffee roasters, such as Nestlé, Kraft, Sara Lee/DE and Procter & Gamble, to pay an 'honest' price for the cultivated coffee beans. By purchasing directly from coffee growers, the intermediate trade can be bypassed and a fair compensation can be paid. The coffee growers form the weakest link in the coffee chain and deserve protection. In the meantime, a few certificates have been developed to ensure that the coffee in question is produced in a responsible manner. This means that a fair compensation is paid to the growers and that there is no damage to the environment. It concerns, among others, the Max Havelaar quality mark, the Utz Kapeh certificate and the Rain Forest Alliance certificate. A part of their coffee is purchased by various coffee roasters in this manner.

The coffee roasters point out the heavy competition, however. If consumers are not prepared to pay the extra costs for the added value of 'honest' coffee, no company can survive. Max Havelaar can allow itself a higher price. It is a niche player and can appeal to idealistic consumers who are willing to pay a bit more for their coffee. Stability of the flavour of the coffee blends also plays an important role. Because the major coffee roasters buy from all over the world and have no admeasurement contracts with certain suppliers, they can ensure a stable flavour. The major coffee roasters serve roughly 10 to 15 % of the world market. With such a scope, you do have to aim at the broad public. Then the stability of the flavour and price play an important role.

The big coffee roasters also state that the low coffee price is ultimately caused by overproduction. Simply said: too much is produced. This is partly due to the end of quota arrangements such as those that applied until the end of the '80s. When the major coffee roasters pay more than the market price, this stimulates even more overproduction. What is the responsibility of the major coffee roasters?

Obesity

Unilever, the fast-food chains and the soft-drink industry are being called to account for making people fat. Obesity is spreading. It is not only a problem in the Western countries; developing countries are also dealing with it. Worldwide there are 1 billion overweight people and 300 million people with serious forms of obesity. Obesity is now one of the leading causes of death. The illnesses caused by obesity cost the Dutch society between 450 million and 2.2 billion Euros per year (Aan de Burgh 2004).

In the discussion, fingers are pointing at the food industry. They try to tempt us to eat hamburgers or ice cream or to drink soft drinks. Are the food concerns actually guilty, or should we blame the consumer? You are what you eat. Unilever top man Anthony Burgmans makes a connection with our altered lifestyle. Calorie intake has hardly increased over the last few decades, but only about half the calories have been burned during that same period. Burgmans: "I blame the electronics industry, because these days people watch television for four, five hours per day. I blame the automobile industry. I also blame the software industry. Thanks to them, children are busy with a PC two, three hours per day. I blame the government, because it is economizing and dropping gym class! Should I go on?"

Chapter 3a

In California the political world has recently joined in the discussion of obesity as well. Here a link is made between the soft-drink industry sponsoring schools and the availability of a certain brand of soft drink at those schools. Thanks to the sponsoring, the schools can buy books and support poverty-stricken parents so that their children can participate in school activities. According to the critics, however, children, who are not very critical, are being influenced without them realizing it. Once accustomed to a certain brand, a child will not easily switch to a different product. This also unconsciously creates the image that soft drinks are healthy. The authority of the school is also reflected on the soft drink that is recommended here.

What is the responsibility of the food industry? Is the increasing obesity of the population the responsibility of the people themselves, or should the soft-drink industry and the fast-food restaurants bear responsibility here? Should restraints be imposed on advertising? Should there even be a tax on high-calorie food, something that has been seriously proposed and is already occurring in some countries?

Striking aspects of the cases

What does the discussion concerning top salaries have to do with the responsibility of the food industry for the obesity of children and the responsibility of the coffee roasters for the low wages of coffee growers? In analysing the cases, a number of matters come to the fore:

- Businesses are pinned with responsibilities that are in fact social responsibilities, in which the company only has a part – sometimes even a very small part. You can call the food industry to account for people becoming fatter, but obesity is initially related to our unhealthy lifestyle. You can blame the coffee roasters for the low coffee price for the coffee growers, while the leeway for the coffee roasters to pay higher prices for the coffee is also related to the willingness of the consumer to pay a higher price for 'responsible' coffee.
- In considering the responsibility of business, people are inclined to reason in 'black and white' terms. The responsibility lies either with the people who are overweight themselves (or the parents of these children), or with the company.
- In the discussions, people readily tend to reduce the complex relationships by singling out a certain connection and understanding it as a causal relationship. The advertising of the food industries, the profits of the coffee roasters at the expense of a fair price for the coffee growers, or the bonuses of the top managers are pointed out as *the* cause of the social problem.
- In the discussions resulting from the described events, values play an important role. The soft-drink industry's drive for profits is in conflict with the health of the youth. The value of the free market is in conflict with a fair reward for the coffee growers. The greed of the top managers is in conflict with a true representation of the reality in the business reports. It is always supposed that *one* social value is placed in a tight corner because (economic) interests are in play that form an impediment for the persons involved for doing justice to the social value. It is reasoned that a choice for the economic interests means a choice against the fundamental social value that is at issue.
- The social debate resulting from these matters is primarily aimed at looking for the guilty parties. When a company or person can be designated as the guilty one, everyone can lean back with a satisfied feeling. It is about the unreliable behaviour of *one* company or *one* person. This approach is understandable. It keeps the issue limited, and therefore manageable. The problem seems to be

solved when we 'remove' the person involved or the company in question goes bankrupt or is totally reorganized. The 'grabby top manager' must disappear, the high-calorie food from the food producers must be kept off the shelves, and the coffee roasters must demand that the suppliers of their coffee pay a fair price. The 'flaw' must be removed. Then we can get back to the order of the day.
- By looking for guilty parties, it is also possible to set aside one's own responsibility. In all three cases, we are dealing with a social problem, whereby we are only too willing to clear ourselves as possible culprits. The concern for a healthy lifestyle is primarily the responsibility of those who consume fast food. The responsibility for the health of the youth lies primarily with the parents. We do not need to change our coffee-buying behaviour. In the example of the top salaries, the public at large is also to blame. At the end of the nineties, the stock market developed into a sort of pyramid game, whereby everyone wanted to profit from the internet hype. Businesses had to show profit gains exceeding 10 %, and preferably with even better profit expectations. That put enormous pressure on businesses and their management. This can be attributed to the bonus-and-option culture.
- If we seek the structural causes in the discussion regarding these affairs, they are generally found in the shortfalls of legislation and the social and internal business control systems[1]. Businesses seek the limits of what is legal and will even exceed them if they are not kept under control via laws and an efficient monitoring system. Therefore, in the debate there is a request to curb the advertising of the soft-drink industry and to levy extra taxes on high-calorie food, and the political world will be pressed to take measures against exorbitant salaries.
- Businesses themselves tend to look for the solution in technology. They are diligently looking for food with fewer calories. Through certificates, the farmers and the intermediary trade must demonstrate that the coffee is produced in a socially and environmentally responsible manner.

The responsibility of businesses for the public interest

The examples fit within a larger picture. In fact, these examples concern the responsibility of businesses to contribute toward solutions for social issues. Looking for guilty parties distracts from the fundamental question facing businesses and, in fact, society as a whole, at this time. Also as regards major social problems, such as climate change, the war against poverty, driving out corruption, the AIDS problem in Africa and, for example, safeguarding biodiversity, businesses play a role. Businesses are called to account on these issues, and various businesses give signals that they experience responsibility in this regard. In the Communist countries, it was the government that was supposed to deal with these issues via a comprehensive control system. Until recently in many Western countries, people relied on a combination of market forces and a corrective government. The Communist system collapsed. But also in the Western countries, where the government had to steer businesses in the right direction via laws and incentive schemes, it seems that the government 'bit off more than it could chew'. Privatization and deregulation resulted in a stronger faith in market forces. Public pressure must stimulate businesses to take responsibility. A consequence of the privatization and deregulation is that social factions no longer turn to the government to steer businesses through legislation. These groups approach

---

[1] In fact, the reasoning here is based on a compliance strategy.

companies directly, or try to pressure businesses by appealing to public opinion. For businesses, it is difficult to work out how to deal with these questions. What is the responsibility of businesses, exactly? Where do the limits of responsibility lie? It cannot be that one company is responsible for all the wrongs in the world. And how can this responsibility be compatible with market-oriented thinking? True, a company needs the trust of the relevant social ranks. In this regard, we can speak of a 'license to operate'. Ultimately, there must be customers who are willing to pay the extra costs of corporate social responsibility, and shareholders must be willing to make risk capital available.

Compliance and market thinking

The compliance view on businesses is in fact seamlessly connected with the market thinking as it has developed over the last centuries in the Western world. In both models, there is actually no room for corporate social responsibility: businesses that, on their own initiative, take responsibility for social issues.

First, let us talk about market thinking. Adam Smith already stated that he has little faith in the good intentions of the entrepreneur. "I have never known much good done by those who affected to trade for the public good" (Of restraints upon the importation from foreign countries of such goods as can be produced at home 1776). He had more trust in the self-interest of the entrepreneur. Via an invisible hand, it would ultimately lead to optimally serving the general interest: "... every individual generally, indeed, neither intends to promote the public interest, nor knows how much he is promoting it ... he intends only his own gain, and he is in this, as in many other cases, led by an invisible hand to promote an end which was not part of his intention. Nor is it always the worse for the society that it was no part of it. By pursuing his own interest he frequently promotes that of the society more effectually than when he really intends to promote it" (Of restraints upon the importation from foreign countries of such goods as can be produced at home 1776).

Market forces ensure that the businesses that can supply their products and services in the most efficient manner will ultimately survive. By using a market model as the basis for finding solutions for the major social problems, consumers and shareholders must in their buying decisions consider the manner in which businesses adopt social responsibility. Moreover, businesses are seen as amoral instruments of the final buyers. The use of certificates and hallmarks (Max Havelaar, Eko hallmark) is based on this line of thinking. The question, however, is whether the knowledge and the insight of consumers can be trusted in the complexity of the connections, and whether it is realistic to expect individual consumers and shareholders to embrace social interests in their buying considerations and investments, and also be able and want to embrace them. In the cases of the harmful effects of frequent consumption of soft drinks or visits to fast-food restaurants, the meagre coffee prices for the coffee growers and the pressure on businesses to show high profit figures, the role of the general public is systematically ignored. Faith in the awareness of the general public (through market forces) falls short, particularly when it concerns unintended side effects of the choices made. Market forces presume that people are driven by their own self-interest, and that what is valuable to the consumer or shareholder is unambiguous, and that you can clearly see a direct relation between the decision to buy and invest and the honoured value.

There is also a second way in which market forces could lead to solving social issues. By incorporating the social costs (externalities) in the price of the product, the price mechanism can contribute toward solving the social problem. One example of this is emission trading. On the basis of the Kyoto protocol, The Netherlands and other countries have committed themselves to bring about reductions in emissions (the main cause of climate change). By setting a price on emissions and making them marketable, one initiates emission trading. No appeal is made to the social responsibility of the consumer, nor to the social responsibility of the entrepreneur. In fact, the system offers room for entrepreneurship. However, in this sense, market forces can only to a limited degree offer a solution for the major social issues. Not all social issues can be expressed as a price or reduced to unequivocal causes on which a price can be set. The problem of obesity is an example of this. The value of health cannot be expressed in a price. But you can show the costs of health care resulting from unhealthy eating habits. What is more, it concerns a combination of causes (too little exercise and the wrong eating habits) in which many parties play a role. In addition, market forces presume an (international) legal system that ensures that the parties will observe the rules of play. The Kyoto protocol and the United States' reaction to it demonstrate the problems involved.

The compliance point of view is based on a negative image of man and organization. Organizations and the people within them tend to pursue their own interests, even at the expense of others. The soft-drink industry's sponsoring of schools is seen as a sophisticated marketing technique. The health of the youth is secondary to the business's drive for profits. The ultimate explanation of the behaviour of businesses is the greed of the top management. Yet this behaviour can be controlled via rules, monitoring compliance and sanctions. The Sarbanes-Oxely legislation and the developments concerning Corporate Governance can be understood on the basis of the prevailing compliance thinking[2].

The compliance approach is linked with the market approach. Like the market model, the compliance approach does not trust the business's own responsibility. Thanks to competition, maximum prosperity is realized. It is tempting to regard steering via (democratically adopted) laws and a (democratically chosen) government-controlled system of laws, rules, procedures and sanctions as a complement to the market forces. Through rules and sanctions, the 'free-rider behaviour' of businesses and the management within them can be prevented. This is how optimal welfare is realized.

A compliance approach cannot offer a solution in the examples described above, however. In order to steer businesses by means of laws, one must be able to understand the problems as simple causal connections, whereby responsibility can be clearly ascribed to certain actors and the effects of certain actions will provide insight beforehand. Moreover, laws, rules and monitoring systems are always running a step behind.

There is also something else more fundamental going on with the compliance approach. The compliance approach eliminates responsibility from businesses and from natural persons – managers, shareholders and consumers. You act responsibly as a citizen when you adhere to the laws, rules and procedures. The compliance approach does not call upon businesses and natural persons to account for their own responsibility. If you want to ensure that the social actors contribute toward solving social issues, then it must be possible to lay down the required behaviour in concrete tasks and roles in a sound manner. A characteristic of the sketched social problems is,

---

[2] Here we are following the distinction that Lynn Sharp Paine (1994) makes between an integrity strategy and a compliance strategy

however, that there are dilemmas. What is required cannot be unequivocally set down. Moreover, there is a certain dynamic. Solutions are found by trial and error. In order to find solutions to major social issues that have been pointed out – such as the problem of obesity and other public-health issues, the problem of climate change, partly due to increasing mobility or the trust in the business world as a condition for economic development – the compliance approach falls short.

Dilemmas, value pluralism and social support [3]

Previously, I had already commented that in the debate on the sketched affairs, the complex connections are often reduced to *one* area of tension between the social or moral values (public health, healthy environment, transparency of the business functions, spreading prosperity, preserving biodiversity, sustainable use of energy sources, etc.) on the one hand and economic interests on the other. The social value is in a tight spot, because economic interests place too much pressure on those involved. Actually, however, there is a conflict between several values. With the problem of obesity, it concerns the consumer's choice, the interests of the schools, for example, the economic interests of the food industry (employment, interests of suppliers, shareholders, authorities) and public health care. In the bookkeeping affairs, it concerns the freedom of top managers to negotiate their terms of employment, the proper functioning of the global economy and a true representation of reality in the business reports. It is not *one* value that is under pressure; it is actually the various values that are all legitimate. Different values have economic sides. Prosperity and the proper functioning of the economy are social values. Just like the other social values, these values strive to have priority over the conflicting values. In fact, we are seeing a cluster of dilemmas, whereby different values are asking for priority.

An important characteristic of the described social issues is that there is no agreement on the best solution. All parties involved define the problem from their own perspective, assess risks in different ways and value the interests at issue in a different way. More accurately: each party involved ascribes a different weight to the values that are at issue. This is why the social issues mentioned cannot be understood merely as technical problems. It also concerns accepting the consequences of certain choices.

However, it also concerns the acceptance of the choices by the people who experience the consequences of the social decisions. I accept a certain nuisance, knowing that major interests are at stake for society. As a child, we lived very close to a railroad. For us, the noise nuisance was a given. You did not even hear the train that thundered by once every half-hour. In fact: when the train did not come one time, you even missed something. We accepted that train and also the growing number of trains on that line. The railway was already there when we came to live there. We realized that rails had to be somewhere. What is more, certain necessary precautions were taken, such as safety measures, to limit the nuisance to a minimum for the people living in the neighbourhood. The encroachment on the value that the rural quietude meant to us was accepted because it served important social values. When businesses consider the stakeholder dialogue, social reporting and the verification of it as a technical solution of the social issues, then they misjudge the nature of these issues. Reducing the calorie content of food or reducing the number of decibels of aeroplane

---

[3] See also Chapter 2 in Kaptein and Wempe (2002)

noise is not enough to achieve social acceptance. I call such an approach the engineer's interpretation of corporate social responsibility.

Corporate social responsibility is aimed at finding widely supported solutions for social issues, with ethics playing a crucial role. Stakeholders will only accept the burdens of social choices when they see the truest possible respect for the values of the victimized stakeholders and a fair consideration of the pluses and minuses for all parties involved.

An appeal to ethics is not sufficient, however. It also concerns an approach that is suitable for understanding the complex social issues. In my view, prevailing ethics do not make the grade in this regard. The classic ethic theories, such as the consequentialist and principalist ethical theories, are all of a monistic nature. They are all based on *one* dominant value or a hierarchy of values on the basis of which behaviour can be judged. In fact, these theories support the technical approach of corporate social responsibility (CSR) that I criticize, and they cannot work with the core of the CSR issue: the plurality of values and the (inter)subjective character of the values that come into play. Several values can be in effect at the same time. When a society makes a decision at a certain moment and thereby chooses for a value that lies at the basis of that interest, this does not mean that a choice is made are against the other values. Moreover, values are always related to the assessments of the persons involved.

Value pluralism rejects the monistic view on moral norms and values. In addition, one must take care not to become caught in the pitfall of moral relativism. According to value pluralism, there are various fundamental values that cannot be reduced to *one* universal value. It is possible that they can clash with each other and that they can even be incompatible. Different values can be in effect at the same time. Sometimes it is possible to indicate priorities. However, it is often not possible to determine what 'the' right answer is. There is not *one* decisive criterion on the basis of which we can ultimately choose. It may not be concluded from the existence of value pluralism that all opinions and the values that lie at the basis of them which the various stakeholders hold are all equally good. The impossibility of indicating priorities can be caused by the fact that our knowledge is not yet sufficient for properly weighing up the values that are in force. Another important reason for the impossibility of indicating priorities is that values cannot be compared.

The moral compromise

Van Willigenburg (2002) recognizes, following Martin Benjamin (1990), the impossibility of determining which values deserve priority, an argument for striving towards a moral compromise. Reaching a moral compromise does not mean that moral values are the object of striking a bargain and that the persons and organizations involved compromise themselves and make concessions on their own integrity. For an airline, the safety of the passengers is one of the most important values. For this reason, most airlines demand a certain number of hours from all the crew members before flying, that no alcohol is consumed and that alcohol tests are conducted. At the same time, the respect for their personnel requires that the airlines assure their privacy. In order to do justice to both values, alcohol tests are usually only conducted on the captains, with the assurance that the test will not be used for other purposes. These airlines accept a certain risk by only subjecting the captains to the test and limiting the use of the test to the purpose for which it is conducted. The pilots accept a certain invasion of their privacy. Reaching a moral compromise can be in the form of negotiation: give and take by the parties involved. According to Van Willigenburg, two requirements must be set within that process of moral negotiation. In order to endorse the autonomy of all those involved, they must endeavour to achieve the most commensurate possible division of pluses and minuses, whereby compromises are made from the different sides. In the example of the alcohol test, all parties compromise. The second requirement of moral compromise is actually at odds with the demand of seeking a proportional distribution. For certain participants in the discussion, some values are so important that they are experienced as determinants of their identity. In seeking a compromise, a more than proportionate weight must be ascribed to such identity-determining values. Within businesses, a discussion can arise about the smoking behaviour of the personnel and the resulting second-hand smoke for the other staff members. For people who are asthmatic, a healthy work environment is so important that a compromise must at least ensure a smoke-free environment for the personnel who are bothered by smoke.

Social surplus value

From a monistic ethics perspective, not being able to choose between values is seen as something negative, as a shortcoming of ethics. However, tensions between values can also form a source of value creation. Value conflicts are then a challenge for those involved to seek better solutions that do more justice to the values there are at issue.

Such tensions are found on all sorts of levels within and around organizations. People fulfil various roles which sometimes involve conflicting expectations. This is intrinsic to life. By working as an employee for a company, the role conflict becomes more complex and weighty. An employee must satisfy a customer, is perhaps a member of the Works Council, wants to treat his colleagues with loyalty but remains the father or mother of a family, has family and friends, and is a member of all sorts of associations. A complex of interests and other role conflicts are the result. Within an organization, different people have different functions. The marketer of a pharmaceutical company is eager to make the most of all the opportunities to put the company's products and services in the limelight. The doctor will be reserved with respect to the marketing techniques and promises that are made in advertising claims. The personnel manager will look primarily at the company's ability to attract new personnel and, for example, have an eye for the personnel's scope to develop their abilities. The security officer worries about information leaks and computer theft. By installing gates, hiring doorkeepers and monitoring the clean-desk policy, he attempts to realize these goals. Various officials are involved in the organization in different ways. Various values lie behind it all. The business deals with a multitude of stakeholders: employees, customers, shareholders, suppliers and, for example, authorities. Each with their own expectations and assessments, and all legitimate. A bank that considers the career opportunities for the employees to be of great importance will have to offer an account manager a new position, also when it means that the customers he serves must then build up a relationship with the new business representative. In society, all sorts of different parties deal with each other. The environmentalists campaign against the encroaching business parks and new residential areas, to the detriment of nature. Businesses want space in order to expand. Local governments must think about tourism, the development of the local economy, housing for their own population and, of course, preserving the natural beauty.

These tensions can be designated as the 'dirty- and many-hands dilemma'. It concerns tensions that occur in the form of dilemmas for managers in daily practice. Together, these three types of dilemmas form the condition under which doing business is possible. As soon as one starts co-operating within a business context and bring products and services onto the market, these three types of dilemmas come about. These are real dilemmas, in the sense that fundamental value conflicts lie behind those tensions. At the same time, within the business context there is often a need for negotiation. Businesses and managers who act on behalf of the business can deal with these tensions in four ways:

They can ignore the value tension. Ultimately, the social (or organizational) field of influence will be the decisive factor for the outcome.

They can try to control the value tension by declaring *one* of the values as the decisive one. This is often a one-sided occurrence, and power is the deciding factor that determines the choice of the value to be maintained.

They can agree on a moral compromise. The moral character of this compromise depends on the questions whether people are striving toward a fair distribution of pluses and minuses and whether account is taken of values that determine identity and the rights of the non-articulate stakeholders.

They can endeavour to rise above the value tension by looking for a situation in which both values can be considered the same time, which will ultimately result in added value for those involved.

With respect to the dilemmas confronting managers in their daily work, it is possible in many cases to look for optimal situations whereby the different values that come into play are thought up together as well as possible. For the process aimed at combining conflicting values, Hampden-Turner and Trompenaars use the term 'reconciliation' (Hampden-Turner and Trompenaars 2000). This 'simultaneous thinking of incomparable or conflicting values' is often possible by breaking through unconsciously and implicitly maintained system limits and by ensuring that we do not only look at the here and now. Questions that bring the dialogue partners on the track of reconciliation are: "Is it possible to pursue one's own values by realizing the values of the other?"; "What is the value in the somewhat longer term?"; and: "What role will the parties involved play in, say, five years?". Reconciliation of the conflicting demands can consist of a straightforward recognition of the impossibility of meeting all the set requirements, expressing remorse, sympathizing with the victim, striving to limit the damage and looking for compensation. It can also entail looking for creative compromises. This approach often means that people allow themselves to be vulnerable, are willing to learn from their mistakes and are open to better ideas.

Transition management

In fact, the described examples require a social turnaround, a transition in society. This involves social changes that can only occur when a great many parties in many places more or less decide at the same time to support one another. The obesity problem makes primary demands of parents regarding their own children, and of the consumer regarding his or her own health. It requires an active contribution from education and health care. And the food industry, the computer, software and electronics companies and the automobile industry must contribute as well. In the first place, it requires awareness of the dilemma and insight into the values that are vying for priority. All parties involved must consider how they can make a contribution toward the solution. Characteristic of such a major social turnaround is that it cannot be realized according to a preconceived plan. It is about many small steps that must be taken in many places, independent of each other, but still more or less on a parallel level. What this transition will look like exactly and where we will wind up is not immediately clear. In this connection, we can use the metaphor of 'the trip south'. We know the direction: to the South, to an honest distribution of pluses and minuses between North and South, toward healthier eating patterns. Where we will wind up is still unknown. Whether it is Barcelona, Rome or Athens is not totally clear. We will have to see whether we must seek the solution in food with fewer calories, better food for babies, more exercise and more self-control by the consumer. We are looking for enthusiastic travelling companions with a feeling for adventure and who want to join us on our way.

Transition management (Rotmans 2003) is the term that is used to explain the process of facilitating such a social turnaround. In fact, all social actors can function as transition manager. The traditional view of the government is that it formulates social goals, which it then attempts to realize using the available policy instruments, such as legislation, taxes and subsidies. Now that the belief in feasible society has been abandoned, it is society (businesses, social groups and the public) that must formulate the concrete goals. This occurs step-by-step, and in many places at the same time. The government, but also businesses or social organizations, can facilitate that process. They can recognize initiatives that go in the desired direction, stimulate parties to 'find each other', help remove obstructions hindering the successful development of these initiatives, and they can make a start possible where financial resources are needed. In concrete terms, it is about looking for likely initiatives aimed at the desired change, for which there is support in the market.

Professional ethics and social changes

In order to find solutions to the sketched social issues, market forces and government regulations fall short of the mark. These issues require a new approach that offers room for value pluralism and relies on social support. In fact, we are looking for a new form of social organization. In addition to the market mechanisms and competition, and besides a corrective and steering government, a new form of social organization will exist, in which there is co-operation between businesses, governments and social organizations. Only through co-operation can the necessary social turnarounds be realized.

This places high demands on the professional. That applies to the manager, the marketer and the researcher. Solutions cannot only be sought in technology. It is quite possible that developing low-calorie food will have an adverse effect. People will think that they can eat even more. A well-intended contribution from businesses aimed at solving the obesity problem can – if the possible adverse effect of the solution becomes visible – be understood as a sophisticated trick from companies aimed at making profits at the expense of people's health. It is possible that setting environmental requirements and social requirements for the coffee middlemen will only wind up helping the big coffee growers at the end of the chain to survive, and that the small growers will be edged out via the competition.

Consequently, the professional will have to be a party in the social debate. He will have to be open to the signals concerning the company, but will also have to share the experienced dilemmas. Only in this way can awareness of the major social issues among all parties involved develop, as well as attunement between all these independent actors, so that the solution to these social problems can be brought closer by, step by step. The professional must develop himself into a transition manager.

References

Aan de Burgh, M., 2004. Globesity: de wereld wordt te dik: wie is verantwoordelijk. *NRC Handelsblad* (23 January 2004).
Benjamin, M., 1990. *Splitting the difference: compromise and integrity in ethics and politics*. University Press of Kansas, Lawrence.
Hampden-Turner, C.M. and Trompenaars, F., 2000. *Building cross-cultural competence: how to create wealth from conflicting values*. Yale University Press, New Haven.
Kaptein, M. and Wempe, J., 2002. *The balanced company: a theory of corporate integrity*. Oxford University Press, Oxford.
Of restraints upon the importation from foreign countries of such goods as can be produced at home, 1776. *In:* Smith, A. ed. *An inquiry into the nature and causes of the wealth of nations. Vol. 4. Of systems of political economy*. London, Chapter 2.
Paine, L.S., 1994. Managing for organizational integrity. *Harvard Business Review*, 72 (2), 106-117.
Rotmans, J., 2003. *Transitiemanagement: sleutel voor een duurzame samenleving*. Van Gorcum, Assen.
Van Willigenburg, T., 2002. Bedreigen morele compromissen onze integriteit? *In: Paper presented during the workshop Integrity in theory and in practice, organized by the Onderzoekschool Ethiek, Vrije Universiteit, Amsterdam, 8 November 2000*.

3b

Comments on Wempe: Conditions for ethical business

Henk Zandvoort[#]

Introduction and overview

In his contribution Wempe addresses the ethical behaviour of business organizations and their agents. He criticizes a reaction to perceived unethical behaviour which he calls the "compliance view", and which he claims is based on a too simplistic analysis of ethical misconduct in organizations and of the underlying social problems that cause the perceived ethical misconduct. He argues that a "plurality of values" lies at the root of many cases of alleged organizational misconduct, and that the 'compliance view' cannot offer solutions to the value conflicts and the social problems which stem from this 'plurality of values'. His aim is to show how business organizations and their agents (professionals, managers) may cope with these value conflicts and social problems and how they may contribute to solving them. In the comments below I will first address the 'compliance view' and what Wempe has to say on that. I will indicate where I agree and also where I disagree. I will then comment on Wempe's views regarding how businesses can deal with the problems arising from conflicting values in society. I will next present my own analysis of a large class of potential ethical misconduct of business organizations, namely cases where business organizations cause environmental damage or other kinds of harm and nuisances which are called 'negative external effects' in the science of economics. Against the background of this exposition, which will include a discussion of Prisoner's Dilemmas, the role of consensus and liability, I will make some final comments on Wempe's view on the social responsibility of business and its agents.

The 'compliance view' of unethical conduct in business organizations

The 'compliance view'

Those adhering to the 'compliance view' of ethical behaviour are trying to reduce ethical misconduct in organizations to the unethical behaviour of a limited number of evil people who need to be warned louder and hit harder in order to make them comply with laws and rules. Hence perceived cases of organizational misconduct are met with a plea for more rules, more control and more punishment. This response disregards the fact that it may not always be clear that executives accused of unethical behaviour actually did break formal rules, and also that there are usually many more involved in the misconduct than just a few key figures at the top. In addition, Wempe claims that the 'compliance view' is inadequate because it does not take into account the plurality of (conflicting) values that is in the background of many cases of alleged

---

[#] Faculty of Technology, Policy and Management, Department of Philosophy, Delft University of Technology, P.O. Box 5015, 2600 GA Delft, The Netherlands. E-mail: h.zandvoort@tbm.tudelft.nl

ethical misconduct of organizations. Because of these conflicting values the issue of what companies should and should not be doing gives rise to ethical dilemmas for businesses and their agents, as conflicting requirements are imposed upon them.

Organizational and institutional failures

I agree with Wempe that reactions to unethical behaviour of organizations are often simplistic and misleading. Often it is suggested that after a few 'bad guys' ('rotten apples') have been removed, all problems have been solved. One may call this the anomaly or incident approach to organizational misconduct. Implicitly or explicitly it is assumed that the contested behaviour was an isolated case, an incident, caused by anomalously behaving individuals. This disregards the possibility of systemic causes for misconduct, either rooted in the organization itself or in the social, legal and political context in which it operates. A similar disregard may be observed in many, notably American-based textbooks on engineering ethics (see Zandvoort, Van de Poel and Brumsen 2000). A related phenomenon can be observed in the analysis of technological accidents. For a long time it was custom to attribute the causes of accidents that could not easily be narrowed down to plain mechanical failures to 'human error'. Such accidents were attributed to e.g. the failure of an operator, who was hence the one to blame. An example would be to lay the exclusive cause and blame for the *Herald of Free Enterprise* disaster with the assistant boatswain who failed to shut the bow doors as he had fallen asleep (for details and background see Van de Poel 2003). Only recently this way of looking at technological accidents has given way to the view that many or perhaps all technological accidents must be considered *organizational* failures. The accident with the *Herald of Free Enterprise* is an example, as the accident could only occur because of persistent inadequacies in the safety procedures and general operation of the ship (for details and a plea for the organizational approach see Dien, Llory and Montmayeul 2003). Organizational failures may ultimately be caused by inadequacies in the social institutions such as law and collective (political) decision making that provide the conditions for organizational behaviour. In this case an even more appropriate term is institutional failures. The analysis of organizational failures pertains to organizational misconduct in general, i.e. including ethical misconduct.

Wempe on conflicting values and the social responsibility of business

Conflicting values and social problems

According to Wempe, the problem that lies at the root of typical cases of (alleged) unethical behaviour of and in business organizations is the plurality of (conflicting) values in society, as this implies that there is typical disagreement on how a certain social problem should be solved and who should do (or abstain from doing) what. Often, Wempe says, genuine dilemmas exist, consisting of fundamental value conflicts. In such cases it is impossible to decide what *the* correct answer to a certain social problem is. I agree that conflicting values frequently occur. Thus, some people think that Schiphol airport should get a fifth (sixth, seventh) runway whereas other people think that this should not be done, and there may even be people who think that Schiphol should be closed down. However, I find it very unclear what Wempe has to say on the methods or procedures that may be used for 'reconciling' such conflicts of values, and hence I fail to see how his recommendations could result into solutions of social problems. Wempe says that the distribution of costs and benefits (of e.g. a fifth, sixth, seventh Schiphol runway?) should be "just" or "as proportional as possible", but he has nothing to say on how this should be determined and by whom. Similarly, he does not clarify what "identity-determining values" are and in which special cases and how "a more than proportionate weight must be ascribed to such identity-determining values". Also, Wempe does not explain how "tensions between values" (that may be related to "fundamental value conflicts") may be "a source of value creation" and what the latter means. In addition, it remains unclear why *the business sector* might be successful in reconciling the conflicting values in society, whereas the political institutions apparently fail. He does not say anything on the role of politics or public choice.

Introduction to the next sections

By discarding the belief in market mechanisms and by discarding the 'compliance view', Wempe also seems to discard the rational actor assumptions underlying economic theory as a basis for regulating the behaviour of private organizations and of persons working in those organizations. I am not ready for that, but I do believe that the legal conditions under which the mechanisms of the market, competition and technological innovation currently operate, make it impossible to expect outcomes that are beneficial or represent progress, in an objective, non-arbitrary way. In the following sections, I will vindicate the claim that a main cause of ethical problems associated with the activities of business organizations is inadequate legislation. As in a democracy legal change must be achieved through the mechanisms or institutions for political decision making, it is here that the 'root cause' of at least some of the problems discussed by Wempe must be localized. My analysis pertains primarily to organizational behaviour relating to environmental damage, sustainability issues, and all (other) kinds of nuisance, harm and risks from technology. This covers a very important portion of the ethical problems relating to business organizations, but I do not claim that all of Wempe's examples can be placed in this category. I will discuss negative external effects stemming from market activities, and how these effects may be internalized into the decisions of actors. I will also analyse these negative external effects using concepts of game theory. Notably, I will introduce the concept of a solution to a Prisoner's Dilemma situation. Like any adequate measure for the internalization of negative external effects, such a solution requires consensus among all involved regarding the rules that govern activities that may affect others. I will argue that stricter liability is a means for securing or at least promoting such consensus, and also that liability should always be strict for activities lacking such consensus.

Negative external effects and the need for consensus regarding market conditions

Invisible hand

The science of economics implies that free markets and competition are to the advantage of all, provided that certain conditions are satisfied. It is to the advantage of the baker to bake bread and sell it to his customers, as it is to the advantage of the customers to buy the bread rather than to bake it themselves. The customers will have more time for other activities (production or consumption) while the baker may enhance the efficiency of baking by exploiting scale, mechanization and other innovations. Everyone involved acts on his or her own interests. There is no organized, centralized co-ordination. Nevertheless the result of all this individual,

Chapter 3b

selfish conduct is beneficial for all, as if led by an invisible hand. Two elements are central to this idea, which was first put forward by Adam Smith in his book *The Wealth of Nations*, published in 1776. The first is that specialization yields efficiency gains. The second is that everyone determines for him/herself which transactions make him/her better off, as every one engages freely into his/her economic transactions. Because of these two elements it is assumed that free markets and competition lead to Pareto improvement, if not for every single market transition, then certainly in the long run. Here, a Pareto improvement is a transition from one situation to another, making at least one person better off and no one worse off.

Negative external effects
One of the conditions that have to be met, according to standard economic theory, is that economic activities should not produce (too many) negative external effects[1]. Textbook examples include air pollution from factories and smoke from cigarettes. Such effects are not accepted by individuals as elements of voluntary transactions, but rather are involuntarily imposed upon them. A more expressive example may be the following. If a Dutch consumer buys a basket of strawberries in a Dutch store, then there is a chance that these were produced in Spain, on land which has been cleared from age-old pine-tree forests[2]. The negative external effects of this activity include: the loss of pine-tree forests; the loss of natural oil resources used for transporting and cooling the strawberries and of freshwater resources used for irrigating the strawberries, and the production of CO_2 contributing to climate change. As free markets and competition spread over the world, so do the negative external effects. Damage to the environment and ecology is taking place on a very large scale, resources are being exhausted, and a variety of obscure but very real and potentially large and far-ranging technological risks is being created. There is no vindication for the claim that the balance is positive; that the result is progress in the unambiguous, non-arbitrary sense of Pareto improvement.

Reducing negative external effects
Hence the following question arises. How can negative external effects be kept within bounds, such that progress, in a non-arbitrary sense, is safeguarded in a system of free markets and competition? The science of economics gives the following answer. First, find ways to internalize the costs associated with these negative external effects in the market prices of the transactions that generate them. An option that has been suggested is to levy an 'eco-tax' upon polluting economic goods or services (see e.g. Baumol and Oates 1988). Another option to be discussed below is to establish liability for damage done (see e.g. Shavell 1987). Second, if the financial translation of the costs of negative external effects which is required for this internalization is for some reason impossible, then the activities causing the negative external effects should be relinquished.

The need for consensus regarding market conditions
How should it be decided *whether* certain costs can be expressed in financial terms, if so, *how*, and if not, which activities may and may not be undertaken? If Pareto improvement is the goal, the only sound answer is this: by *consensus of all involved*. Here, those involved include not merely those who take actively part in the

---
[1] There are other conditions as well that I will not go into here.
[2] Source H. Piek, Nederlandse Vereniging tot Behoud van Natuurmonumenten, private communication.

economic activities, but also all those who may experience the negative external effects. If such consensus is lacking, then there is no guarantee whatsoever that free markets and competition will lead to Pareto improvement (for more on the need of consensus among all involved regarding measures such as 'eco-taxes' intended to internalize negative external effects of market activities see Zandvoort in prep.).

Prisoner's Dilemmas and their solutions

The tragedy of the commons

It is important to note that even if it is assumed that individuals enter into those economic activities that, given the circumstances, are for the best of their own interest, this is no guarantee whatsoever that the net result, including negative external effects, will be (experienced as) positive. The possibility to the contrary is illuminated by the 'tragedy of the commons'. This 'tragedy'[3] was intended by its author (Hardin 1968) as an analogy for the environmental problems of present-day society. It involves the following. A community has common grazing land where every farmer is allowed to let his sheep graze. As every farmer wants to improve his personal standards of living, each tries to put as many sheep as possible out to graze there. Above a certain limit, for every extra sheep there will be less grass and therefore a lower yield per sheep. Ultimately even the total revenue will decrease: the grass disappears and the sheep grow thin. Everyone would hence benefit from a 'grazing quota' (dividing up the pasture is another possibility) but agreements on that do not develop spontaneously. For each separate farmer it is disadvantageous to invest time in that or to limit the number of sheep he places in the meadow when all the others are spending all their energy on expanding their flocks. The downfall of every farmer thus seems to be unavoidable.

Prisoner's Dilemma situations and game theory

Similar situations occur frequently in present-day society. They are at the basis of many ethical problems related to technology. Such situations have become known as 'Prisoner's Dilemma situations'. A Prisoner's Dilemma (PD) situation involving many persons may be characterized as follows (Zandvoort 2003):
- Everyone would be better off, at least in the long run, if all people were to place certain restrictions on their actions.
- For each separate individual imposing these self-limitations is disadvantageous compared to those who do not subject themselves to the same limitations. Moreover, the contribution to the total negative effect of one or several non-conformers is negligible.

PD situations have been extensively studied in game theory. Game theory is sometimes described as the mathematical analysis of strategic interaction between people. Strategic interaction is defined as situations where the decisions (and the subsequent actions) of two or more individuals together determine the outcome. Game theory was created by Von Neumann and Morgenstern in their book *Theory of games and economic behavior* (1947). Since then, its concepts and results have been used to model many actually occurring situations of strategic interaction. Some of these concepts and results are used below.

---

[3] If a tragedy is defined as an *inevitable* course of negative events, as many seem to do, then 'tragedy' is actually not a correct word for the processes that I deal with in the text.

Two-person PD games

The simplest PD situation is a two-person one-off game, where there are only two actors or 'players' who meet each other only once. The situation is often further simplified by assuming that the actors have two options each (called strategies in game-theoretic language). A PD 'game' of this type is defined by the following so-called 'pay-off matrix' that specifies how the outcome for each depends upon the actions of both (see e.g. McLean 1987, chapter 7):

		You	
		Co-operate	Defect
I	Co-operate	3, 3 R, r	0, 5 S, t
	Defect	5, 0 T, s	1, 1 P, p

Figure 1. The pay-off matrix for a two-person PD game

Here, 0, 5 should be read: the outcome of the combination {I co-operate, You defect} is valued 0 by Me, and 5 by You. (Initially, the words co-operate and defect are mere labels for the two strategies that both players have available, but below I will vindicate for repeated PD games the connotations of these words in ordinary language.) The game is defined solely in terms of how the players *order* their valuations of the four possible outcomes, not on quantitative values. Hence, if the outcomes for Me of the four possibilities {we both co-operate}, {I co-operate, You defect}, {I defect, You co-operate}, and {we both defect} are called R, S, T and P, then a game is called a PD game if I order these outcomes as $T > R > P > S$ ($X > Y$ means that I prefer X above Y), and if You order the corresponding outcomes for You as $t > r > p > s$. Hence, any change in the numbers 3, 0, 5 and 1 which retains their ordering renders again a PD game, as does multiplying My (or Your) values with an arbitrary positive constant.

Equilibrium strategy for PD games

Suppose I want to choose the strategy that optimizes My outcome for this one-off game. Then I must consider two possibilities: either You co-operate, in which case it is best for Me to defect; or You defect, in which case it is also best for Me to defect. Hence in both cases My optimal result ensues by defecting. The same reasoning applies to You, and hence if You and I are economical, selfish players, outcome 1, 1 will materialize. This may even remain true for a repeated PD game, which is a situation where the same players encounter each other repeatedly (or continuously) in a PD situation. The tragedy of the commons exemplifies this.

Solution for a repeated PD game

However, for a repeated PD game there is an unambiguous, non-arbitrary solution available, namely a contractual agreement that ties both players to the strategy 'co-operate'. This solution is non-arbitrary or objective in the sense that it leads both players to results that are better than if there were no agreement, for outcome R, r (or rather a series of such outcomes) is valued higher by each player than a series of outcomes P, p. Hence, for repeated PD games, the strategy labelled 'co-operate' may be associated with socially optimal behaviour, and likewise to 'defect' with socially sub-optimal (or detrimental) behaviour. The non-arbitrariness or objectivity of these definitions rests on the notion of Pareto improvement: socially beneficial behaviour is defined as behaviour that is beneficial for all. (Clearly, if there are negative external effects for parties other than the players, then the agreement should be extended to those others as well.)

Nature and costs of solution

The described solution requires that players reach an agreement on certain rules. That requires organization and effort, that is: costs ('transaction costs' in terms of economic theory). It is the task of politics to design such rules and to secure agreement on these rules (for an introduction to this view on the role of government see Mueller 1989, esp. Chapter 1). If there are many actors involved in the agreement, a system of enforcement will be needed as part of the agreement: an arbiter/controller with the authority to levy sanctions to violators of the agreement. Also such a system of enforcement involves additional costs. (It might be an element of the agreement that these costs be recovered from the sanctions, as these costs would not have to be made if there were no violators.) The costs for settling and executing the arrangement should not exceed the collective gain from the arrangement if it is really to be a solution to the PD situation. This is the case if these costs do not exceed (R-P) + (r-p) in our simplified model.

PD situations are omnipresent

In our technological society, characterized by environmental pollution, ecological damage, depletion of natural resources and all kinds of technological risks, Prisoner's Dilemma situations are omnipresent. The human production of CO_2 and its impact upon the climate may be one of many examples. It is very relevant to know whether a certain problem has the traits of a Prisoner's Dilemma situation, for in that case a solution of the type described above is in principle available: a set of shared rules coupled to a system of control and sanctions for violation on which there is agreement. If such solutions are available but nevertheless not realized, this may be blamed on failing political institutions or procedures of public choice, rather than on conflicting values or unsolvable moral dilemmas.

Legal systems and consensus

The solution to PD situations described above amounts to what is commonly known as a legal system. It is important to notice that such systems, including sanctions for violations, can in principle be based on the consensus of all involved. More to the point, such systems *must* be based on that consensus in order to represent a solution to a PD situation in the sense described above. In contrast to this, in the present societies consensus is not required and does not exist concerning the legal conditions that govern economic and technological activities. National political decision making proceeds on the basis of majority rule at best. The qualification 'at best' refers to serious problems with representation in the contemporary democracies. See Mueller (1989) for vindication. In addition, present democracies continually sanction activities with significant actual or possible negative effects for people living outside those democracies, without having secured their consent[4].

The role of liability

The result regarding the need for consensus on market conditions that was reached earlier above can be obtained more directly and more generally from the ethical principle of restricted liberty, also known as the no-harm principle. In addition, by invoking the reciprocity principle governing situations in which this principle has been violated, the analysis can be complemented with considerations regarding liability. This renders a 'default rule' for liability in the absence of consensus on market conditions. These two issues are the subject of the present section, which draws on work of J.F.C. van Velsen (2000; 2003).

Restricted liberty

The restricted-liberty principle can be stated as follows: *Everyone is free to do what he/she pleases as long as he/she does not harm others.* An equivalent formulation of the restricted-liberty principle is the right to be safeguarded: *Everyone has the right to be safeguarded from the consequences of another person's actions.* An implication is that actions are allowed if and only if either there are no (possible) consequences for others; or those who will experience the (possible) consequences have consented after having been fully informed.

Reciprocity and liability

The principle of reciprocity specifies how violations to the restricted-liberty principle may be dealt with: *He who violates a right of another one will be reacted to in a reciprocal way. That means that somebody who infringes a certain right of another will himself lose that same right insofar as that is necessary (and no more than that) in order to correct the original violation or to compensate for it and in order to, if necessary, prevent further infringement.* Reciprocity implies that anyone not respecting another person's right to be safeguarded and thereby causing another person harm, loses his own right to be safeguarded, in the sense that he may be forced to repair or compensate the damage. Hence the reciprocity principle implies strict, that is, non-conditional liability for activities that lack the prior consent of those who may experience the consequences. In actual reality this is very often not the case (a concise history of legal liability for technological/commercial activities is contained in Zandvoort 2000).

The role of liability

The consent required for activities that may affect others may be provided by a general rule such as a law on which there is consensus among those involved. In order to secure that consensus, it may prove necessary that actors accept liability for

---

[4] It is sometimes claimed that between states, there is consensus decision-making regarding market conditions. An example that is sometimes quoted is the WTO. But others have stressed that in fact the situation is not one of real, that is, free consent Steger, M.B., 2002. *Globalism: the new market ideology.* Rowman & Littlefield, Lanham. . Nevertheless there seems to be agreement regarding the *desirability* of consensus on the international level regarding market conditions. It is remarkable how few show awareness of the discrepancy of this with that other idea, adopted by very many either explicitly or implicitly but almost always in an entirely uncritical way, namely that *within* nations majority rule would be an adequate procedure for making collective decisions.

negative consequences of their contemplated actions. However, if such consent is lacking, liability should always be strict, in the sense of non-conditional.

Recapitulation of the above analysis

A large class of cases of alleged ethical misconduct of businesses relates to negative external effects generated by business activities. If adequate measures for internalizing such negative external effects into the behaviour of actors will not be implemented, such allegations of ethical misconduct of businesses cannot be expected ever to stop. Such measures require the consensus of all involved, i.e. of all those who may experience negative external effects from the activities considered. The same result was obtained from an analysis in terms of Prisoner's Dilemmas. Many situations regarding environmental pollution, the depletion of natural resources and the creation of technological risks apparently have the structure of Prisoner's Dilemmas. Such dilemmas can be solved in a non-arbitrary way, by introducing a set of (legal) rules that have the consent of all involved. If a problem has the structure of a Prisoner's Dilemma, then such solutions exist. If they are nevertheless not identified and implemented, this should be blamed to failing political institutions and procedures (such as majority rule) rather than to conflicting values. The analysis of (internalizing) negative external effects and of (solving) Prisoner's Dilemmas are both based on the concept of Pareto improvement, and on the claim that social progress can be objectively defined only as Pareto improvement. An implication of this is that, if an activity embodies social progress, then that activity can obtain the consent of all who are affected by it. Finally, I have argued, on the basis of the ethical principles of restricted liberty and reciprocity, that liability should be strict for all activities that do not have the consent of all those who may be affected.

The social responsibility of business agents

Wempe's aim is to advise business and its agents as to how they can resolve the value conflicts in society that are at the basis of ethical problems relating to the activities of business organizations. As was explained at the beginning of my comments, I have two main points of criticism regarding his advice. The first point is that I find the methods and procedures described by Wempe very unclear. Hence it remains unclear how and why these methods and procedures can perform their intended functions. I find Wempe's exposition of little help regarding the question: how should situations be dealt with in which people's actions affect other people and where they do not agree on the norms or values on which the actions or their consequences are evaluated? The second point is this. One way in which I can make sense of the proposed methods and procedures for resolving value conflicts in society is that they are an attempt at a(n) (ill-defined) method for creating solutions to PD situations, in the sense explained above. But such a method, even if it would be sound in itself, cannot be executed by business organizations, contrary to what Wempe suggests. Instead, adequate procedures and institutions of public choice and law are needed. I have argued (1) that adequate procedures of political decision making should be based on consensus rather than on majority rule as is presently the case, (2) that the introduction of stricter liability may be a means of securing consent to activities which otherwise would not be consented to, and (3) that as long as actual political decision making is not based on consensus, liability should be strict. Against this background, I want to end with the following remarks regarding ethical business and what Wempe has to say on that.

The role of business organizations in solving PD situations

Even if business organizations cannot perform the functions of politics and government, it remains a relevant question what the role of business organizations and their agents in political processes should be. In view of the above analysis one would expect that business organizations and their agents who want to perform their activities in an ethical way would actively contribute to the realization of the changes that are required according to that analysis, notably the introduction of liability laws that are more strict than at present, and of political decision making that is more than at present based on consensus. At the very least, one would expect critical and open-minded assessments of the inadequacies of the actual political and legal systems. In practice, very little of this can be observed. This leads me to the following remarks with which I will end my commentary.

Freedom of speech of persons associated to (business) organizations

This freedom of speech is very limited in the organizations that populate the present societies (see Zandvoort in press, for substantiation and for implications). This holds both for business organizations and for government organizations. This effectively keeps many people's mouths shut on a large range of very important issues, including the need for stricter liability laws discussed above. It appears to me that ethical business organizations without the freedom of speech of those working there is almost a *contradictio in terminis*. Unfortunately, Wempe's contribution does not address this subject.

The political responsibility of business agents

The limited freedom of speech mentioned above adds to the ethical responsibility that those who legally speak and decide for business organizations carry for the political outcomes of what they say and do and of what they fail to say and do. Again, Wempe's contribution has nothing to say on the responsibility of business agents (professionals, managers) considered from this perspective.

References

Baumol, W.J. and Oates, W.E., 1988. *The theory of environmental policy*. 2nd edn. Cambridge University Press, Cambridge.
Dien, Y., Llory, M. and Montmayeul, R., 2003. Investigation on the methodology of organisational accidents and lessons learned. *In: Pre-proceedings of the 24th ESReDA seminar on safety investigation of accidents, 12-13 May 2003, Petten, The Netherlands*. JRC/Institute of Energy, Petten, 177-185.
Hardin, G., 1968. The tragedy of the commons. *Science,* 162 (3859), 1243-1248.
McLean, I., 1987. *Public choice: an introduction*. Blackwell, Oxford.
Mueller, D.C., 1989. *Public choice II*. Cambridge University Press, Cambridge.
Shavell, S., 1987. *Economic analysis of accident law*. Harvard University Press, Cambridge.
Steger, M.B., 2002. *Globalism: the new market ideology*. Rowman & Littlefield, Lanham.
Van de Poel, I.R., 2003. Resonsibility within organisations. *In:* Zandvoort, H., Van de Poel, I.R., Brumsen, M., et al. eds. *Ethics and engineering (syllabus)*. Faculty

of Technology, Policy and Management, Delft University of Technology, Delft, Chapter 6.

Van Velsen, J.F.C., 2000. Relativity, universality and peaceful coexistence. *Archiv für Rechts- und Sozialphilosophie,* 86 (1), 88-108.

Van Velsen, J.F.C., 2003. *Het recht van de logica: voorwaarden voor een vreedzame samenleving.* Eburon, Delft.

Von Neumann, J. and Morgenstern, O., 1947. *Theory of games and economic behavior.* 2nd edn. Princeton University Press, Princeton.

Zandvoort, H., 2000. Controlling technology through law: the role of legal liability. *In:* Brandt, D. and Cernetic, J. eds. *7th IFAC symposium on automated systems based on human skill, Joint Design of Technology and Organisation, June 15-17, 2000, Aachen, Germany, preprints.* VDI/VDE-Gesellschaft Mess- und Automatisierungstechnik (GMA), Düsseldorf, 247-250.

Zandvoort, H., 2003. Responsible conduct of organisations and the role of law. *In:* Zandvoort, H., Van de Poel, I.R., Brumsen, M., et al. eds. *Ethics and engineering (syllabus).* Faculty of Technology, Policy and Management, Delft University of Technology, Delft, Chapter 7.

Zandvoort, H., in prep. Globalisation, environmental costs, and progress: the role of consensus and liability. *Water Science and Technology.*

Zandvoort, H., in press. Good engineers need good laws. In: *Proceedings of the 31st SEFI Annual Conference, Porto, September 7-10, 2003.*

Zandvoort, H., Van de Poel, I. and Brumsen, M., 2000. Ethics in the engineering curricula: topics, trends and challenges for the future. *European Journal of Engineering Education,* 25 (4), 291-302.

Zandvoort, H., Van de Poel, I.R., Brumsen, M., et al. (eds.), 2003. *Ethics and engineering (syllabus).* Faculty of Technology, Policy and Management, Delft University of Technology, Delft.

RESPONSIBLE AUTHORSHIP AND COMMUNICATION

4a

The responsible conduct of research, including responsible authorship and publication practices

Ruth Ellen Bulger[#]

Responsible conduct of research

When attempting to identify the norms for the responsible conduct in biomedical research, it is important to identify areas in which scientists have come to some agreement on what are accepted norms and areas in which such consensus has not been reached. It is also important to understand the principles that underlie such norms. Just as the Belmont Report (Department of Health Education and Welfare 1979) provided three guiding principles for research involving human participants (respect for persons, beneficence and justice), the principles underlying biomedical research ethics also need be to elucidated.

In an attempt to move toward defining these underlying principles for the responsible conduct of research, Bulger (1994) has suggested a possible way to classify responsibilities according to four guiding principles. She proposes the first principle as a constellation of values including honesty, integrity, truthfulness and objectivity in the way that scientists plan, execute, record, interpret and publish their work. These values have been uniformly singled out as the key to the doing of science well (National Academy of Sciences 1992). As scientists strive for objectivity, they benefit by the examination of their intellectual biases and elimination of conflicts of interest.

Second, is the way scientists show respect for the other, including the humane care and use of animal subjects, respect for human participants in clinical research, for students and other research collaborators, and for the environment. Scientists show respect for students and their collaborators by sharing data, products and information with them freely, and by the proper attribution of credit to them for their ideas and work. Demonstrating collegiality is important and yet difficult in the modern research setting (Bulger and Bulger 1992). Respect for the environment is shown by undertaking only important research so resources are not wasted. It is also demonstrated by thorough literature searches to prevent the useless repetition of work that has already been done.

The third principle relates to the competence of the trained investigators in obtaining and transmitting their research data. This includes using valid techniques and proper statistical evaluation. The results of one's study should be promptly published so that others can benefit from the fact that they were done.

Finally, the stewardship of society's resources relates to how scientists ply their trade and choose problems to be studied. Since much health-related research is funded by society, the scientist has an ethical responsibility to demonstrate good stewardship

---

[#] Department of Anatomy, Physiology, and Genetics, Uniformed Services University of the Health Sciences, 4301 Jones Bridge Road, Bethesda, Maryland, 20814 USA. E-mail: Rbulger@usuhs.mil

of the resources that are provided. Exactly how this responsibility is expressed is an area of disagreement among scientists. For some, it includes only a commitment to do the science responsibly and well. This is justified by the fact that it is hard to know just what basic science may become important to future progress. For others, it would include choosing a research topic of importance to society and its members. A further commitment would be to help identify and participate in the resolution of ethical quandaries uncovered by the science that the scientist produces (Reiser and Bulger 1997).

Although agreed-upon norms have been defined in some areas of scientific endeavour, the majority of situations that the scientist must address lie in grey areas that remain undefined, murky, with many pros and cons on how to proceed and little agreement among scientists as to a uniform solution (Jasanoff 1993).

In response to well-publicized incidents of misconduct in science, the National Institutes of Health, the principal source of funding for biomedical sciences in the U.S., required instruction in several areas of the responsible conduct of research for all fellows supported by National Research Service Award institutional training grants. Although the means by which this education was to be provided was not specified (e.g., classes, lectures or mentoring), the areas to be covered were conflicts of interest, authorship and publication, misconduct and data management (National Institutes of Health 1989; 1990).

In 2000, a more extensive instructional mandate was put forth by the Office of Research Integrity (ORI) of the Department of Health and Human Services, which required instruction in nine areas of the responsible conduct of research (RCR). The policy required instruction in the following: acquisition, management, sharing and ownership of data; mentor/trainee relationships; responsible authorship and publication practices; peer review and the use of privileged information; collaborative science; human-volunteer research; humane care and use of animals; research misconduct; and conflicts of interest and commitments. The instruction was to be for all "staff at the institution who have direct and substantive involvement in proposing, performing, reviewing, or reporting research, or who receive research training supported by PHS funds or who otherwise work on the PHS-supported research project even if the individual does not receive PHS support." Although this guidance was later suspended, some institutions have continued to move toward this type of instruction for various individuals involved with scientific research (US Department of Health and Human Services: Office of Research Integrity 2000).

Finally, in light of several instances of problems with research studies at Universities that involved human participants, the Office of Human Research Protections (OHRP) stopped human participants research at several major universities. OHRP subsequently has required that all investigators dealing with research involving human volunteers receive education before being allowed to do this type of research.

In response to ethical problems in biomedicine in the U.S., there have been calls for increased education and accreditation of investigators and administrators as well as increased audits to ensure that regulations are being met within the institutions in which the research is being done. Yet it is important to realize that creating and enforcing regulations provides a minimum level for ethical behaviour. Scientists must not be creating a culture of regulation, but a culture of conscience. In looking toward the future of the ethics movement in the biological and health sciences, Reiser (2002) reminds us that "This next phase of development in the biohealth sciences will produce new discoveries, but some will be of a different sort than those to which biohealth scientists are accustomed. Until this time biological scientists have single-mindedly explored the environment of nature. They must now turn their attention to the environment of their profession and focus their vision inward, on themselves".

Ethics of responsible authorship and publication

One of the nine areas in which the Office of Research Integrity required education was the ethics of authorship and publication practices. That is the area that I have been asked to address in more detail for this conference. Communication of research results is a central element in the doing of science; in fact, there is little reason to do scientific research if the results are not shared with others in the community. In addition, authorship with subsequent publication leads to several important outcomes for the individual scientist's career including the assignment of credit as well as responsibility for the research, the recording of the accomplishment as a measure of one's scientific performance, allowing the work to be repeated by others and thereby validated, and placing the work in perspective with other research already published in a way that allows scientists to build on the work of others.

Publication of research results in journals is the way that results have been recorded for centuries, and it remains the major way that scientists communicate. It relies on the ability of scientists to trust the work reported by others. As Steven Shapin (1995), sociologist of science, so aptly states, "It needs to be understood that trust is a condition for having the body of knowledge currently called science…To suggest that scepticism and distrust should be very much more common in science is, in effect, to take the position that much of our modern structure of scientific knowledge should be unwound, put into reverse, and ultimately dismantled. Instead of laboratories for the production of new knowledge, we should build great facilities for the close reinspection of what is currently taken to be knowledge. Grants will be given for checking routine findings: published reports will look more and more like laboratory notebooks: libraries will have to be expanded to house an unimaginable vast literature reporting upon acts of distrust: relations between scientists will become uncoordinated, unproductive, and unpleasant".

Even though scientific communication is of paramount importance, the present environment for authorship and publication is in a rapid state of flux. This is partly due to the changes in the way authorship is being defined, as well as to the marked and continued increase in research funding and the subsequent increase in the number of articles written and journals to publish them. In addition, there has been a marked growth of the impact of electronic resources being used by scientists to communicate their work. The laborious hand searches of published literature (with subsequent reprint collection) have been replaced by the almost effortless electronic literature searches with on-line access to many published abstracts and manuscripts. Major changes are occurring in the way scientists handle information that is published and even more drastic, even paradigmatic, changes are promised for the upcoming years.

Who are authors?

Profound changes in criteria for authorship have occurred during the last couple of decades and how the new authorship criteria are applied across laboratories. In past years, a scientist who has authored 500-1000 or more biomedical articles was greatly admired and often chosen for influential positions such as department chair. Yet in a 30- to 40-year career publishing that many papers would mean the publication of 1-3 different papers per month over the entire time span. How much effort such a person expended per manuscript was never questioned. In the present environment, publishing an entire intellectual piece of work is being encouraged, while publishing numerous small papers – previously referred to as the Least Publishable Unit (LPU) (Broad 1981) or the practicing of salami science (Huth 1986) – is criticized. In addition, the publication of the same material more than once (repetitive publication) is wasteful. In fact, if similar material is to be published in two places, it must be referenced to avoid self-plagiarism.

The definition of who deserves the title of authorship is being narrowed. The International Committee of Medical Journal Editors (2001) has published Uniform Requirements for Manuscripts Submitted to Biomedical Journals (www.icjme.org). Its published definition of authorship has been accepted as a standard by over 500 medical journals. The definition states that authorship should be based on three conditions: "1) substantial contributions to conception and design, or acquisition of data, or analysis and interpretation of data; 2) drafting the article or revising it critically for important intellectual content, and 3) final approval of the version to be published".

In addition the report states that "All persons designated as authors should qualify for authorship and all those who qualify should be listed". This is a strong statement eliminating both honorary (including those who do not meet the criteria for authorship) and ghost authorship (omitting anyone who does qualify for authorship). Yet Flanagin et al. (1998) have shown that these practices are still occurring in journals.

Along with authorship credit goes the responsibility for the work. The ICJME policy states: "Each author should have participated sufficiently in the work to take public responsibility for appropriate portions of the content. One or more authors should take responsibility for the integrity of the work as a whole, from inception to published article". Such a policy underlies the growing trend for journals to require authors to state the specific role of each author in the work. Standard practices would then advocate using the acknowledgments section, not author status, to give credit for those who have only provided various resources or help with the work, such as advice or manuscript review. In fact, Rennie, Yank and Emanuel (1997) have suggested that the term 'author' be replaced by the two categories of 'contributor' and 'guarantor', clearly indicating the role of each. They believe that such a system would be precise, understandable and fair, and would discourage misconduct. Although the suggestion of Rennie, Yank and Emanuel (1997) has not been accepted, the practice of listing the specific role of each author is expanding and partially fulfils their aim.

Self-regulation by journals

In light of the urgency in dealing with terrorism, including the harmful use of infectious agents by terrorists, two national meetings were held in the US in January 2003 that included scientists, publishers, security experts and government officials. The topic of the meeting was how journals and meetings of scientific societies could handle new scientific information both responsibly and effectively when safety and security issues raised by submitted papers could be exploited by terrorists and therefore should not be published. The group concluded that potential harm of publication could outweigh potential societal benefits. In such a case, the submitted paper should be modified or not published. Journals and scientific societies could encourage scientists to communicate this type of research results in other ways that maximize public benefit while minimizing risk of misuse (Journal Editors and Authors Group 2003).

The informatics revolution in authorship and publishing

The access to references and to published materials directly via the Internet or by library subscriptions allowing on-line access to reference databases and the published literature are revolutionizing how scientists use the literature available after about 1966. Internet access to resources provides rapid new ways to search the literature and therefore can increase the productivity of scientists. Internet access to published literature is available to scientists who lack extensive library or financial resources beyond a computer and Internet access. This is often the case for scientists in the developing world.

Medline (or comparable) access for identifying or checking references has become the standard for modern scientists and reviewers. Some journals have even made their entire publications available electronically, either to subscribers or to the general public. In 1999, Varmus proposed a broad new two-tiered initiative for improved access to the original manuscripts and to publications (see Marshall 1999b). The access to some of this previously published material is now becoming a reality as PubMed Central, a central repository containing a body of literature in the life sciences that can be easily searched on the Internet. Journals are being encouraged to distribute their publications in PubMed Central after a short (1-6-month) delay. Markowitz reports that some scientists are being encouraged not to submit manuscripts or to review manuscripts of others for journals not releasing their contents to PubMed Central (Marshall 1999a; Markowitz 2000).

In light of the rising costs of journal subscriptions and worldwide acceptance of the Internet as a valid publication medium, Markowitz (2000) proposed that scientists re-examine the current paradigm for publishing research according to present journal-publishing policy. He points out that scientific authors turn their copyright over to journals without any financial rewards. In fact, authors also may pay page charges or purchase reprints and their institutions must pay for subscriptions to these journals. In addition, scientists not only provide the articles to the journal but also provide free review services. If the authors were to retain copyright to their scholarly manuscripts, they could publish them on the Internet either with or without prior journal publication. Markowitz points out that if manuscripts were to be published directly on the Internet, then some type of peer-review system might need to be developed, perhaps paid for by the authors, their institutions or the commercial advertising presently being used to fund similar activities in journals (Markowitz 2000). However, an alternate kind of review system could be developed similar to that presently employed by Amazon.com and a growing number of retail on-line businesses, in which those reading the books or purchasing the products do a post-purchase review giving their assessment of the value of the purchase that is displayed on-line with the specifics of the book or product.

Markowitz (2000) sees many advantages to this type of web-based system. They include having rapid access to the contents, built-in cross-referenced hyperlinks, integrated searching, inclusion of original data, multimedia formats, less expensive than journal publication, more environmentally correct, available wherever one has computer access, and available to those without large financial backing. He points out that a similar freely accessible self-publication policy exists in the field of physics sponsored by the American Physical Society in co-operation with the Los Alamos

Laboratory (American Physical Society 2002). It is possible that such profound changes in how scientists communicate may lead to problems for scientific journals. In fact, publishing in journals as we know it might disappear, and be replaced only by electronic communication (Markovitz 2000). Although it is always difficult to undergo such a profound change in behaviour as such a change to publication on the Internet involves, the advantages to such a system must at least be considered.

Publication with a broader definition of scholarship

Authorship of original articles describing scientific research is still the coin of the realm in science. More recently, however, there has been a development of a broader definition of what scholarship entails including areas other than that limited to traditional scientific discovery. New ways of documenting scholarship besides journal publishing accompanies this movement. Such ways include the creation of a teaching portfolio. These ideas have been influenced by the perceptive book by Ernest Boyer (1990) titled "Scholarship Reconsidered: Priorities of the Professoriate", in which he argued for a fuller range of scholarship. Boyer stressed the importance of creative new directions not only in the *scholarship of discovery* (increasing the stock of human knowledge), but the disciplined, investigative efforts of the *scholarship of integration* (giving meaning to isolated facts), *the scholarship of application* (the synthesis of traditions of academic life), and the *scholarship of teaching* (not just presenting the material, but transforming and extending it). Changes in academic promotion policies, at least at our University, are being affected by this new definition of scholarship. A more inclusive dynamic view of scholarship would continue to involve communication/publication of what has been learned, but would be enhanced by the use of broader multimedia forms of communication including video and audio presentations and artistic renditions. The review of the scholarship could be accomplished either before or after posting by input from the experts and consumers of the material.

Challenges lie ahead as scholars seek to balance the advantages of the tried and true means of journal publication with the possibilities becoming available for a vastly increased audience of computer-literate individuals.

(The opinions expressed in this article are those of the author and do not reflect opinions of the Department of Defense of the United States or of the Uniformed Services University of the Health Sciences, Bethesda, Maryland, USA.)

References

American Physical Society, 2002. *Transfer of copyright agreement.* [http://forms.aps.org/author/copytrnsfr.asc]
Boyer, E.L., 1990. *Scholarship reconsidered: priorities of the professoriate.* The Carnegie Foundation for the Advancement of Teaching, Princeton.
Broad, W.J., 1981. The publishing game: getting more for less. *Science,* 211, 1137-1139.
Bulger, R.E., 1994. Toward a statement of the principles underlying responsible conduct biomedical research. *Academic Medicine,* 69 (2), 102-107.
Bulger, R.J. and Bulger, R.E., 1992. Obstacles to collegiality in the academic health center. *Bulletin of the New York Academy of Medicine,* 68 (2), 303-307.

Department of Health Education and Welfare, 1979. *The Belmont report: ethical principles and guidelines for the protection of human subjects of research.* U.S. Government Printing Office, Washington DC.

Flanagin, A., Carey, L.A., Fontanarosa, P.B., et al., 1998. Prevalence of articles with honorary authors and ghost authors in peer-reviewed medical journals. *JAMA: the Journal of the American Medical Association,* 280 (3), 222-224.

Huth, E.J., 1986. Irresponsible authorship and wasteful publication. *Annals of Internal Medicine,* 104 (2), 257-259.

International Committee of Medical Journal Editors, 2001. *Uniform requirements for manuscripts submitted to biomedical journals: writing and editing for biomedical publication.* ICMJE Secretariat office, American College of Physicians, Philadelphia. [http://www.icmje.org/]

Jasanoff, S., 1993. Innovation and integrity in biomedical research. *Academic Medicine,* 68 (9 Suppl), S91-S95.

Journal Editors and Authors Group, 2003. Statement on scientific publication and security. *Science,* 299 (5610), 1149.

Markovitz, B.P., 2000. Biomedicine's electronic publishing paradigm shift: copyright policy and PubMed Central. *Journal of the American Medical Informatics Association,* 7 (3), 222-229. [http://www.pubmedcentral.nih.gov/picrender.fcgi?artid=61424&action=stream&blobtype=pdf]

Marshall, E., 1999a. E-biomed morphs to E-biosci: focus shifts to reviewed papers. *Science,* 285 (5429), 810-811 (as cited by Markowitz 2000).

Marshall, E., 1999b. Varmus circulates proposal for NIH-backed online venture. *Science,* 284 (5415), 718.

National Academy of Sciences, COSEPUP, 1992. *Responsible science. Volume 1. Ensuring the integrity of the research process.* National Academy Press, Washington DC.

National Institutes of Health, 1989. Requirement for programs on the responsible conduct of research in National Research Service Award Institutional Training Programs. *NIH Guide for Grants and Contracts,* 18 (45). [http://grants1.nih.gov/grants/guide/historical/1989_12_22_Vol_18_No_45.pdf]

National Institutes of Health, 1990. Requirement for programs on the responsible conduct of research in National Research Service Award Institutional Training Programs. *NIH Guide for Grants and Contracts,* 19 (30). [http://grants1.nih.gov/grants/guide/historical/1990_08_17_Vol_19_No_30.pdf]

Reiser, S.J., 2002. The ethics movement in the biological and health sciences: a new voyage of discovery. *In:* Bulger, R.E., Heitman, E. and Reiser, S.J. eds. *The ethical dimensions of the biological and health sciences.* 2nd edn. Cambridge University Press, Cambridge, 3-18.

Reiser, S.J. and Bulger, R.E., 1997. The social responsibilities of biological scientists. *Science and Engineering Ethics,* 3 (2), 137-143.

Rennie, D., Yank, V. and Emanuel, L., 1997. When authorship fails: a proposal to make contributors accountable. *JAMA: the Journal of the American Medical Association,* 278 (7), 579-585.

Shapin, S., 1995. Trust, honesty, and the authority of science. *In:* Bulger, R.E., Bobby, E.M. and Fineberg, H.V. eds. *Society's choices: social and ethical decision making in biomedicine.* Committee on the Social and Ethical Impacts of Developments in Biomedicine, Institute of Medicine, National Academy Press, Washington DC, 388-408.

US Department of Health and Human Services: Office of Research Integrity, 2000. *PHS policy on instruction in the responsible conduct of research (RCR).* Office of Research Integrity ORI, Washington DC. [http://ori.dhhs.gov/html/programs/finalpolicy.asp]

4b

Comments on Bulger: The responsible conduct of research, including responsible authorship and publication practices

Henk van den Belt[#]

For a sociologically informed research ethics

The starting point of *research ethics* can be found in Robert K. Merton's classic essay from 1942, "The Normative Structure of Science" (Merton 1973). This essay is usually taken as a key contribution to functionalist sociology of science, but I would argue that it also holds relevance for the new field of research ethics. In the essay the American sociologist described and codified the so-called *'scientific ethos'* – the normative framework for the conduct of science – as consisting of a series of norms and values that could be summed up in the acronym CUDOS: communism, universalism, disinterestedness and organized scepticism. To avoid misunderstanding: the norm of 'communism' implies that the results of scientific research are published within a reasonable time and assigned to the community: "Secrecy is the antithesis of this norm; full and open communication its enactment" (Merton 1973, p. 274). Merton wrote his essay long before universities in the USA and elsewhere would be thoroughly commercialized: filing patents was still frowned upon. But Merton was keenly aware that the scientific enterprise was not just co-operation but also competition (he used the phrase 'competitive co-operation'). The competition was not so much, or not primarily, about the acquisition of monetary rewards, but turned on recognition and esteem, credit and reputation. Nevertheless, as frequent priority disputes testified, the struggle could be very intense.

In her paper Professor Bulger writes: "Authorship of original articles describing scientific research is still the coin of the realm of science". Merton said something similar, but with a slightly different accent: "Honorific recognition by fellow-scientists is the coin of the scientific realm". So it is not authorship *per se*, but the credit that it may earn among one's colleagues, that in Merton's view is the currency of science. He sees science as a social system in which original contributions, of course after having been published and thus made available to the community, are exchanged for recognition and esteem and all the possible rewards that may go with these. Such rewards range from *eponymy* (e.g. Boyle's law, Brownian movement, Mendelian genetics), medals, fellowships and membership of prestigious organizations to ennoblement (in the UK) and, of course, the Nobel Prize. A modern quantitative measure of 'recognition and esteem' is provided by the Science Citation Index.

The reason why I recall Merton's work on the sociology of science here is that I want to argue that *research ethics* needs an injection of sociological realism (or better sociological scepticism) to become less naive and also much less anodyne in its

---

[#] Applied Philosophy Group, Wageningen University and Research Centre, Hollandseweg 1, 6706 KN Wageningen, The Netherlands. E-mail: HenkvandenBelt@wur.nl

M.J.J.A.A. Korthals and R.J. Bogers (eds.), Ethics for Life Scientists, 63-66.
© 2004 *Springer. Printed in the Netherlands.*

prescriptions. Merton made one assumption that seems immensely reasonable to me, to wit, that scientists are just like normal mortals and do not possess special moral qualities; they are *not* "recruited from the ranks of those who exhibit an unusual degree of moral integrity" (Merton 1973, p. 276). At the very least, this seems a good 'null hypothesis' to start with. Merton also remarked – remember: back in 1942! – that fraud and deceit are extremely rare, virtually absent, in the annals of science. For Merton, the only way to explain this fact (if it is a fact) was to point out that "scientific research is under the exacting scrutiny of fellow experts" and that "the activities of scientists are subject to rigorous policing", thus strengthening the effect of a successful internalization of the scientific ethos (Merton 1973, p. 276). Since Merton wrote his path-breaking essay, the practice of science – or at least our views of science – must have changed dramatically. From the late 1970s on, the science journals and the mass media have been busily reporting an unending series of affairs involving fraud, deceit, plagiarism and other forms of 'misconduct', especially in the biomedical sciences. In fact this trend has also been the main factor behind the rise of *research ethics*, at first in the USA and later elsewhere. The question is whether research ethics can offer useful solutions to do something about the problem. I am somewhat doubtful.

Will the mandatory instruction in research ethics for those who receive grants, as demanded by the National Institutes of Health (NIH) and the Office of Research Integrity (ORI), really help to reduce future incidents of misconduct, especially in hot areas of science where competition is intense, commercial pressure heavy and the stakes are high? Will the pedagogical effort to impart the values of 'honesty', 'integrity', 'truthfulness' and 'objectivity' on PhD students and young researchers be effective in promoting decent behaviour on their part? Or will this simplistic pedagogy to instill ethical 'correctness' perhaps more likely evoke rebellion? (To me 'environmental correctness' sounds almost as bad as 'political correctness'; if I have to show respect for the environment by undertaking only 'important' research so resources are not wasted, I am strongly inclined to say: "To hell with the environment!".)

In this connection I want to draw attention to a problem which Professor Bulger also obliquely broaches. Referring to Sheila Jasanoff, she writes: "Although agreed-upon norms have been defined in some areas of scientific endeavour, the majority of situations that the scientist must address lie in grey areas that remain undefined, murky, with many pros and cons on how to proceed and little agreement among scientists as to a uniform solution". This could give the impression that we have, on the one hand, domains governed by agreed-upon norms where it is clear for everyone how to proceed, and, on the other hand, grey areas where everything is murky. This would be the wrong impression, I think. One major point of criticism of Merton's sociology of science by the newer, 'post-Mertonian' and 'social-constructivist' school in science studies (to which Sheila Jasanoff also belongs) is precisely that general norms (even if agreed upon) are *always* to some extent negotiable and open to multiple interpretation when they have to be applied to concrete situations.

Take for example Professor Bulger's remark (relating to the third principle of competence): "The results of one's study should be *promptly* published so that others can benefit from the fact that they were done" (italics mine; this could be seen as an echo of Merton's description of the norm of 'communism' – 'full and open communication' – see the quotation above). The critical question, when it comes to implementing this norm, would of course be: How 'promptly' exactly is 'promptly'? The sociologist of science Michael Mulkay provides a telling example. When the radio-astronomy group at Cambridge, UK, published the first paper on the newly discovered pulsars in 1968, rival groups accused them of having unduly delayed publication of the relevant data. The Cambridge group, however, defended their publication policy. They denied that there had been any 'undue' delay or unjustified secrecy at all. After all, they argued, researchers need time to check their results in order to publish high-quality work! Thus this episode shows that there may be strong disagreement among scientists about what exactly a general norm like the norm of prompt publication implies in a concrete situation (Mulkay 1976).

Towards a culture of accountability?

One way to overcome the problem of the interpretative flexibility of general norms, at least partially, is to install special bureaucratic agencies that are charged with overseeing the rules and norms that are to be followed. As part of their mandated mission, such agencies will undertake to clarify and more strictly define the pertinent norms. However, the forceful attempt to ensure that researchers and their institutions comply with a set of imposed formal rules could easily lead to a 'culture of regulation', or a 'culture of accountability', as the British philosopher Onora O'Neill would call it (O'Neill 2002a). To me, even the name 'Office of Research Integrity' (ORI) already sounds quite intimidating and a little bit Orwellian – something like the 'Ministry of Truth'. In practice, however, the ORI does not seem to have real bite, as scientists already resist its attempt to gather information on such low-key unethical behaviour as authors citing papers they haven't read and condemn this as the agency's meddling in areas beyond its purview (Soft responses to misconduct 2002). A more ominous example is provided by the Danish counterpart of ORI, the Danish Committees on Scientific Dishonesty, which in January 2003 condemned Bjørn Lomborg's controversial book *The Skeptical Environmentalist* as "objectively speaking, deemed to fall within the concept of scientific dishonesty" (Abbott 2003). I find it rather disquieting that a research-ethics panel apparently feels no scruples to pass judgment on a lively controversy involving a lot of scientific and political issues. In my view, this is not the way to settle such a debate.

Professor Bulger alludes to the "calls for increased education and accreditation of investigators and administrators as well as increased audits to ensure that regulations are being met within the institutions in which the research is being done". I fully sympathize with her remark that "Scientists must not be creating a culture of regulation, but a culture of conscience". Still I am concerned that *research ethics* 'as-it-really-exists' may become increasingly institutionalized as part of a growing culture of accountability. Onora O'Neill gives a very disturbing account of the 'audit explosion' and the growth of a 'culture of accountability' in all social domains beyond the original financial context, especially in the public sector of the scientific, health and service professions (O'Neill 2002a; see also O'Neill 2002b). The 'new wave of audit' makes environments 'auditable'; "audits do as much to construct definitions of quality and performance as to monitor them"; they produce a "drift to managing by numbers"; "the construction of auditable environments has necessitated record keeping demands that serve only the audit process" (quotations from Michael Power in O'Neill 2002a, p. 132-133). The new culture of accountability may enforce 'trustworthy' behaviour, but it does not breed trust – it rather breeds suspicion! The proliferation of distrust finally raises the question: Who audits the auditors? "Ultimately there is a regress of mistrust in which the performances of auditors and inspectors are themselves subjected to audit" (Michael Power, quoted in O'Neill

2002a, p. 133). A second route to trustworthiness, according to O'Neill, has aimed to construct a more open public culture, by ensuring that information is available to the public. She refers to the Committee on Standards in Public Life (installed by John Major in 1995 and chaired by Lord Nolan), which promulgated seven ethical principles for the conduct of office holders: selflessness, integrity, objectivity, accountability, openness, honesty and leadership (colloquially known as 'the Nolan Principles'). Since the Nolan reports, new requirements have been widely implemented: "Office holders are required to act only in the public interest, to be open, to avoid conflicts of interest and to declare any interest (the standard for identifying a declarable interest is that others would perceive it as such); declarations of interest are made public [...]" (O'Neill 2002a, p. 135-136). The unexamined and questionable assumption of this quest for openness and transparency, according to O'Neill, is that by enforcing trustworthiness among office holders the new culture will actually receive more public trust. The intended field of application of this openness offensive is much broader than the so-called 'ethics movement' in the biological and health sciences, but is its thrust so much different?

References

Abbott, A., 2003. Ethics panel attacks environment book. *Nature,* 421 (6920), 201.

Merton, R.K., 1973. The normative structure of science (essay orig. publ. 1942). *In:* Storer, N.W. ed. *The sociology of science.* The University of Chicago Press, Chicago, 267-278.

Mulkay, M.J., 1976. Norms and ideology in science. *Social Science Information,* 15 (4/5), 637-656.

O'Neill, O., 2002a. *Autonomy and trust in bioethics.* Cambridge University Press, Cambridge.

O'Neill, O., 2002b. *A question of trust.* BBC Reith Lectures 2002. [http://www.bbc.co.uk/radio4/reith2002/]

Soft responses to misconduct, 2002. *Nature,* 420 (6913), 253.

5a

Professional ethics and scholarly communication

Hub Zwart[#]

Introduction: a short history of scholarly reading and writing

Science (or, more generally, the procurement, propagation and dissemination of knowledge)[1] has not always been a profession based on the transmission of written materials. On the contrary, when Western science originated in ancient Greece some twenty-five centuries ago, dissemination usually took the form of verbal exchange and personal tuition (Zwart 2001). In the context of schools, scientists would share their (more or less secret) knowledge with a limited number of pupils by means of immediate verbal interaction. In those days, the transmission of knowledge was not a writing profession. The use of texts implied the risk of losing control over one's ideas. Moreover, the use of textual information was not really necessary. The emphasis was on principles and method. The teachings conveyed by teachers to their disciples basically consisted of a limited set of ideas that could be transmitted and elucidated orally, while brief expositions of principles would alternate with sessions of exercise. This applied to disciplines like philosophy and mathematics, and more generally to the scientific practices of investigation that emerged within the ancient schools of Physicists and Pythagoreans.

The situation changed, however, when a new generation of teachers, the Sophists, stepped forward. They initiated a different kind of investigation called 'history' (Zeller 1980). Besides history in the present sense, this new branch of learning included ethnographical, sociological and geographical studies as well. A different kind of knowledge was produced, namely *factual* knowledge. And the wealth of materials collected by generations of 'historians' had to be filed somehow, in order to prevent their loss, and thus the use of texts became important. Sophists took great care, moreover, in developing their rhetorical skills, both as orators and as writers. This was due to the fact that for Sophists knowledge was a commodity to sell and the way it was presented affected in a substantial way its (practical) relevance. Some of these 'stylistics of knowledge' can still be found in the way modern scientists disseminate their results.

The significant difference between philosophy in a strict sense (as the 'unwritten' science of principles) and sophistry (as the art of producing elegant prose) is still noticeable in the case of Plato. While in his written dialogues he entered into competition with the Sophists, notably in terms of literary style and narrative composition, he abstained from the use of textual materials as a teacher in the context of his own school (Wippern 1972). This had to do with the 'propagandistic' character of his dialogues. They were aimed at publicly refuting the Sophists' positions. Indeed,

---

[#] Faculty of Science, Mathematics and Computer Science, Department of Philosophy and Science Studies, University of Nijmegen, Postbus 9010, 6500 GL Nijmegen, The Netherlands. E-mail: haezwart@sci.kun.nl

the difference between these two scholarly styles is still noticeable today, even if the motivations of modern scholars and ancient philosophers may not always coincide. Prominent scientists still have the habit of practicing two genres. On the one hand, they will write compact articles directed at a limited number of fellow experts. Usually, this part of their output is utterly unreadable and incomprehensible for lay audiences. But every now and then they will have recourse to plain and elegant prose as well, in order to produce pieces of text meant to address much broader audiences. A very well-known contemporary example is physicist Stephen Hawking. His *Brief History of Time* uses a jargon and a style of argumentation that would probably be totally unacceptable if submitted to a scientific publisher, but as an educational text it has substantial merit. It succeeded, for example, in fascinating lay audiences with mind-boggling concepts like evaporating black holes and other exotic objects. Also among philosophers, important examples of this double writing practice can be given. Immanuel Kant, for example, had been a successful writer in his youth, addressing relatively broad audiences, using elegant and quite readable prose during the first part of his career, before switching to the abstract, reader-unfriendly, 'academic' and critical work later in life – the work that *really* made him famous (Vorländer 1977) – although he continued to write newspaper articles (on issues such as 'Enlightenment') as well. Likewise, Leibniz produced a series of essays on popular subjects during his lifetime, written for audiences interested in philosophy and science but lacking systematic training, while in the meanwhile he accumulated the bulk of his scientific output in unpublished notes and manuscripts. We will come back to this interesting phenomenon later on.

What is important is that 'real' scientists and scholars, from the early onset, took a rather ambivalent stance towards writing as such, and to publishing in particular. The knowledge of principles directly appeals to the pupil's own intellectual faculties. It is impersonal and concise. The shift to written discourse, with all the stylistic and rhetorical devices that come into play, makes the attention of these pupils (readers) shift from what they see and think for themselves to 'information', and finally to the 'opinions' of authors – who in the long run will become 'authorities'. In scientific *prose*, we encounter the scholar *as a person*, expounding his personal opinions and trying to persuade his audience to accept his particular point of view, for example by showing off his erudition, his acquaintance with literary sources – thus forcing his readership into a more or less passive and receptive role. For indeed, although initially 'historians' (collectors of opinions and facts) would base their knowledge on personal observation, gradually the *reading* and processing of textual materials became increasingly important. Knowledge claims were increasingly based on textual sources produced by previous generations of researchers. Thus the writing scientist became a *compiler*. The archetypical representative of this branch of knowledge is without doubt Plinius; whose gigantic compilation called *Historia Naturalis* was really a storehouse of factual knowledge. In his works, the emphasis is not on principles, not on 'thinking', not on method, but on the techniques for storage and retrieval of tremendous amounts of information.

Thus, whereas initially science relied on thinking (i.e., the contemplation of basic principles), the emphasis gradually shifted to reading, processing and interpreting texts. Scholars became professional readers and commentators rather than thinkers. Or, to put it more adequately, thinking and reading became intimately interconnected. Reading became the professional scholarly activity *par excellence.*

During the Renaissance, this association or even identification of scientific scholarship with reading was vehemently reinforced[2], but during the subsequent

Scientific Revolution, both activities were disconnected once again. The new scientific professional ideal as it emerged during the early modern period demanded that scientists, rather than relying on textual sources, should themselves become active contributors to the production of factual knowledge. But these facts had to be produced in a *methodological* manner. An experimental set-up was not simply a device that enabled a researcher to assemble interesting facts, but rather a tool for posing and answering very specific questions. And it was this new scientific method that increasingly distinguished scientific (i.e., reliable) facts from anecdotal (unreliable) information (often borrowed from textual sources). The realization of the unreliability of written materials was actually one of the causes that helped the scientific method to come about; as an example, the case of Copernicus stands out. He made some wrong predictions about planetary motion and realized later that this was due to the fact that he had used data 'collected' by ancient astronomers. These turned out to have been doctored in order to fit their preconceived cosmological and cosmogonical theories. Later generations of astronomers realized the importance of collecting data in a methodological and systematic manner.

The new type of professional scientific activity (systematic observation, using sophisticated equipment, and notably experimentation) entailed the emergence of a new scientific genre: the research paper (as well as the scientific journal: a periodic compilation of research papers). During the initial stage, research reports were extensive and often rather personal. They contained exciting coverage of the vicissitudes of doing research. The focus would be on the 'context of discovery'. Activities and findings were reported in a more or less 'narrative' fashion. Gradually, however, scientific discourse became less personal, less subject-dependent. Scientific articles submitted to journals had to be written and composed in accordance with standard formats. These formats increasingly determined the way in which a scholarly contribution was to be framed and phrased. And not only the composition of an article was formalized, also the terms and sentences available to professional authors were normalized and standardized. The author as a person gradually disappeared from the text. Thus, whereas in ancient Greece the individuality of the author gradually became more important, now we noticed a trend in the opposite direction: from personal to impersonal. The author as a person became less important, because methodology and principles were regarded as more important than personal prestige. The development of a personal style was no longer appreciated. The text became, to a certain extent, anonymous and depersonalized. All written materials adhered to one and the same professional style. The article no longer described the 'context of discovery' of scientific investigations. On the contrary, detours, failures, anecdotes, hesitations, quotations and other discursive frills were increasingly removed from the final publication. The context of justification increasingly determined the structure of the article. Instead of describing the actual course – the 'history' – of the investigation, the published article reflected professional methodological standards. The publication emphasized the methodological *principle*s the author subscribed to. Only those findings and considerations that reflected the basic experimental script (question, hypothesis, procedure, measurements, statistical analysis etc.) were presented. The final result of this development was the scholarly article as we know it today: compact, using standard terms, formulae and phrases, preferring quantitative information (usually presented by means of matrices or tables) to more extended narrative accounts. The research as it is published reflects the methodological norm rather than personal experiences. And this has had an impact on laboratory life as well. The methodological requirement of reproducibility, for example, made real experiments increasingly successful in meeting the standard protocols delineated by the ways in which published experiments were being described. For indeed, whenever we use standard formats to describe what we do, such formats will tend to 'realize' themselves in the long run. They will influence ('normalize') our practice. The way we describe our experiments has an impact on the way we perform them.

Besides *methodological normativity*, another form of normativity has had a tremendous influence on scientific discourse, namely *ethical normativity*. The moral ideal of the unprejudiced, self-critical, reliable, unimpeachable, accurate, autonomous, responsible researcher is reflected in the self-presentation of the author of a scientific text. The moral principle of fairness, moreover, requires from scholars that they cite or quote the authors whose data or ideas they consciously used. At the same time, they justify the way they perform their professional work by referring to the examples, methods, results and statements of others. Therefore, the act of citing one another, both as an act of fairness and as a way of justifying one's own activities and decisions, is an important element in scholarly writing. But there is a moral import in 'strictly' methodological decisions as well. As standard formats force us to be brief, it is sometimes difficult to decide what precisely should be revealed, and what can safely be left out. How to select relevant information from the profuse archives of laboratory experiments? These questions are not always easy to answer.

Some problems involved in scholarly writing today

I started this article with a brief exposition of historical developments because I firmly believe that today, science as a profession (and scholarly authorship as a practice in particular) is once again experiencing a period of intense change. Basic principles and concepts, basic methodological and ethical conceptions are called into question, and this involves moral and conceptual uncertainties. Scientists are confronted with a growing number of normative issues and ethical dilemmas concerning publication, authorship, reliability of information, as well as with questions that used to be of interest to philosophers and social scientists only, such as: What is an author? What is a journal? What is a publication? Although some of these issues have a long history, their gravity has increased of late, due to the growth of scientific data and the acceleration of the process of data dissemination, stimulated by the emergence of new communication and information technologies. Scholars can now, more easily than before, circulate works in various stages of completion. Moreover, the economic conditions for scientific work have changed. There was a time when the bulk of scientific research was conducted by members of the leisure class who had ample time and means at their disposal to pursue their scientific interests. Nowadays, however, because of the transformation of science into a networked activity in which competition for results plays a fundamental role, scientists seem to be forced to worry continuously about funding and to behave as entrepreneurs, acting in a calculating and strategic manner. In this article, some issues of professional ethics will be addressed that have to do with the 'telos' of scientific research: the act of publishing one's research results. I will discuss commercialization and intellectual autonomy; the impact of the informational revolution; the problem of pluralism of ethical styles; and the fairness or unfairness of citation practices.

Commercialization and intellectual autonomy

"The late twentieth century has witnessed a scientific gold rush of astonishing proportions: the headlong and furious haste to commercialize genetic engineering. This enterprise has proceeded so rapidly – with so little outside commentary – that its dimensions and implications are hardly understood at all... Biotechnology research is now carried out in more than two thousand laboratories in America alone. Five hundred corporations spend five billion dollars a year on this technology... The commercialization of molecular biology is the most stunning ethical event in the history of science, and it has happened with astonishing speed. For four hundred years, science has always proceeded as a free and open inquiry into the workings of nature... Scientists have always rebelled against secrecy in research, and have even frowned on the idea of patenting their discoveries, seeing themselves as working for the benefit of mankind. When, in 1953, two young researchers in England, James Watson and Francis Crick, deciphered the structure of DNA, their work was hailed as a triumph of the centuries-old quest to understand the universe in a scientific way. It was confidentially expected that their discovery would be selflessly extended to the greater benefit of mankind. Yet that did not happen. Thirty years later, nearly all of Watson and Crick's scientific colleagues were engaged in another sort of enterprise entirely. Research in molecular biology has become a vast, multibillion-dollar commercial undertaking... Suddenly it seemed as if everyone wanted to become rich...".

The passage quoted above is taken from Michael Crichton's bestseller *Jurassic Park* (1991). Those who are familiar with the history of science will agree that Crichton's account is not very adequate from a purely historical point of view and reflects a widespread and idealized picture of science and scientists. It is not true that for centuries scientists could afford to work in a completely selfless manner until they suddenly awoke from their altruistic slumber a few decades ago, due to the commercialization of biotechnology. Rarely were the conditions of discovery for scientists as optimal as Crichton suggests. Be this as it may, the passage is interesting enough if we read it, not as a historical account, but rather as a kind of sermon in which an ethical ideal is sketched and the contours of a professional ethic for scientists is fleshed out, precisely by presenting the reverse image of such an ideal (that is, by presenting the idealized past as a critical mirror that allows us to reflect on some of the problems involved in current conditions). According to Crichton, commercialization constitutes a threat, while selflessness and altruism are virtues – and he presents this view as something that speaks for itself and should be taken for granted. At present, all of a sudden, the world of science is out of joint. Modern science has lost its moral innocence and splendour.

Although from a historical point of view Crichton's account is clearly an exaggeration, we at the same time have to confess that it is not *completely* untrue. Until recently, scientific research was conducted in two different, more or less clearly separated realms: namely in a commercial and in an academic environment. These two contexts corresponded with two different styles of doing research, with two different professional and ethical orientations. In a commercial setting, knowledge was produced by companies that tried to convert their knowledge products into patents, while in a scholarly setting knowledge was produced by scientists who respected the intrinsic value of scientific knowledge and were satisfied with having their achievements recorded in scientific journals. The best way to reward them and honour them for their achievements was by citing them. At present, the boundaries between these two different settings tend to blur, even though the magnitude of this phenomenon (as should be expected) varies greatly according to which discipline we take into account. We will come back to this discipline specificity later. This leads to ethical uncertainties, such as: is a scientist a scholar or rather an entrepreneur?

One of the consequences is that university researchers are becoming increasingly dependent on commercial organizations and companies, not only for funding of their research, but also for tools, equipment, samples, research opportunities and expertise. These companies will often formulate constraints that are at odds, to a greater or lesser extent, with the traditional mores of science. And as a rule, the value conflicts that evolve from this situation will become especially pertinent when the moment has arrived to publish one's research results. This should not come as a surprise, since the act of publishing as such involves already some relevant moral decisions. One of those, as noted in the introduction, is the selection of reliable and relevant results. Does commercialization endanger the intellectual autonomy of scientists in this and other kinds of considerations? In some cases, for example, researchers are only allowed to publish positive results. In other cases, the company wants to have a say in whether the manuscript should be submitted for publication at all or not. Recently, a PhD student was faced with the following situation. She had ordered a sample from a patented product, produced by a private company X, owned by a former university professor and specialized in producing biomaterials for medical purposes. Besides funding part of her research, this company also provided her with research tools (highly specialized software). She eagerly awaited the delivery of the package. Upon arrival of the sample she discovered that a legal document was added to it, indicating that manuscripts have to be submitted to and approved by the company before they can be published or otherwise disseminated through scientific channels. . This document contained the following lines:

"Prior to publication of any work relating to the material, investigator shall furnish X with copies of any such proposed presentation at least thirty days in advance of the submission of same in order to allow X to comment on same. Further, investigator shall acknowledge X's contribution of said material, unless specifically instructed to the contrary by X".

Moreover, it was pointed out in a detailed manner what the legal consequences would be for herself, her department and her university, should she fail to comply with the instructions stipulated in the document. It seems like the intellectual and moral autonomy of the scientist is being seriously threatened.

Scholarly communication in the information age

Problems connected with commercialization are not the only difficulties awaiting those who decide to enter an academic research career. Scientific research, and especially scientific authorship, may entail a number of other quandaries as well. To begin with, our ideas about authorship have evolved under material conditions that were rather different from the ones we are confronted with now. A number of changes may be listed:
- The practice of multiple authorship (i.e. the growing number of authors and co-authors of scientific publications). This increases the opportunity for scientific collaboration, but can also have the unwanted collateral effect of decreasing the sense of moral responsibility of the individual scientists involved for the published results.

- New tools for intellectual exchange (such as electronic preprints made available on the Internet) blur the distinction between (informal) communication and publication.
- Globalization and the dramatic increase of the number of papers submitted for publication.
- The acceleration of knowledge production. Due to this development, the time lag between submission and publication of materials will be increasingly experienced as detrimental to the need of spreading innovative results in a short time. Authors will look for alternative channels to disseminate results, notably to peers.

A number of key concepts are called into question by these and similar developments. For example: what is an author? Or rather: who should count as author? Is only the first author the 'real' author of a scientific article? Should the head of the department be listed as author, for example because he provided the funding as well as the infrastructure needed to carry out the research? And what about the technician who did the laboratory work? Or the colleague who gave advice on statistical issues? Can a clear boundary be drawn between 'authors' and those contributors whose names are mentioned under the heading 'acknowledgements'? Is it morally acceptable that scientists or research groups agree to cite one another mutually in order to increase their number of hits on the citation index? Difficult decisions have to be made and conventions may prove to be rather unstable.

On authorship

Is a manager an author? In 1897, the Russian physiologist Ivan Pavlov (1849-1936) published his *Lectures on the work of the main digestive glands*. Because of this influential achievement, he was nominated for the Nobel Prize. There were, however, important reservations. To what extent were Pavlov's works really Pavlov's? At that time, physiological research was migrating from scientific workshops (where experienced researchers were performing small-scale experiments themselves, together with a limited number of co-workers or assistants) to real laboratories that involved division of labour and made it possible for researchers to become managers and to leave the bulk of the laboratory work to their co-workers. The nominee had himself pronounced his *Lectures* "the deed of an entire laboratory" and credited his co-workers by name for conducting the experiments on which it was based. Was his work an original contribution to science, or rather a compilation of a series of experimental dissertations by others (Todes 2002, p. xiii)? This is an early example of a problem that is still with us today. Two years ago Craig Venter et al. (2001) published his famous account on the structure of the human genome. In this article, the names of 285 authors were listed. What are the implications of this development for our ideas on authorship? Traditional concepts no longer seem to reflect the way research is actually done.

What is an author? Michel Foucault (1995) pointed out that in the course of history, scientific authorship has performed a number of different functions. In the medieval era, for example, the name of the author – the name of *Aristotle* for example – was regarded as a guarantee of truth, as a reliability indicator. In modern times, the name of the author came to be used to designate methods, tests, scales, special numbers, diseases, bodily parts, comets, or straits and islands – a phenomenon which is known as eponymy (cf. Merton and Storer 1973). Nowadays, authorship has lost much of its former prestige and is used in a rather technical manner. We use the name of the author for the retrieval of documents and for the evaluation or monitoring (by means of a citation index) of individual scholars, research groups and research institutes. Still, contemporary scientists and scholars are not radically consistent in following this trend. The new, technical function of authorship still has to compete with other functions. Something of the old prestige of authorship is still left intact; we still value the papers authored by the 'big names' as almost intrinsically better that those written by less famous scientists. This reveals an interesting tension between the trend towards anonymization of the author, versus the aura some authors still seem to have, an aura which allows them to function as 'authorities', and their name as indices for reliability and quality.

And what is a publication? In a laboratory, a stream of data is gradually being processed from raw materials to printable products. But when exactly is it time to stop gathering data and to submit research findings for publication, notably in the case of pressing competition with rival research groups? Due to the time gap between submission and publication of output in scientific journals, scientists will look for alternative, electronic channels of dissemination. The high-energy-physics experiment on CPT invariance conducted by Walter Oelert and his team at CERN, resulting in the discovery of antimatter, may stand as an example here. While publication of their results in *Physics Letters* was delayed due to methodological problems noticed by one of the referees, the news that antimatter was discovered spread via the Internet, forcing CERN to take the unusual step of issuing a press release on a scientific result before the scientific paper had been published (Fraser 2000). Once a paper is published, however, the process of refinement and updating may continue. What version should count as 'final' when scientists are able to update continuously the materials they made electronically available? Indeed, electronic dissemination nowadays allows scientists to enter into an interactive exchange process with their readers. What does 'final' mean when authors are able to replace existing electronic versions by revised ones, or to solicit and incorporate feedback? Electronic publishing, as an alternative to hard-copy formats, allows the reader to play a more active role in the production of scientific output. It is clear that concepts like 'authorship' and 'publication' have to be redefined and their moral implications reconsidered.

The future of the library

Over the centuries, scientists and scholars have produced and accumulated enormous quantities of textual materials. This immediately brings us to the question of the archive: the storage and retrieval of scholarly writing. During the epoch when research and reading were intimately associated, even in the natural sciences, as explained above, the library would constitute the centre of scholarly life. The famous library of Alexandria, together with those libraries as existed in medieval monasteries, still constitutes the basic image, the archetypal idea so to speak of what a library is. An archetype, however, not only entails a basic image, but also a basic scenario, a script. An archetypal library is a collection of documents so overwhelmingly large that we feel discouraged and intimidated by it. The story, however, is likely to end with the library being destroyed by a fire – either in real life (such as the library of Alexandria) or in fiction (such as the libraries in the writings of Eco and Borges for example). And there is an element of grief involved in such an event, of course, but also of relief. The basic attitude of scholars towards libraries (or towards archives in general) is an ambiguous one. On the one hand, our work depends on their existence. On the other hand, we want to be liberated from such a heavy burden: the net result of centuries of intellectual labour. We want to stand on our own feet, rather than on the shoulders of paper 'giants'. This ambivalent attitude parallels and is intimately connected with the relation already analysed between textual (compilative) and empirical knowledge.

From a philosophical point of view, however, the 'invisible fires' that occasionally destroy complete libraries are much more interesting than the real ones. Every now and then in the academic world, an Exodus occurs. A new generation of scholars 'decides' to leave the library and to opt for a journey through the desert, producing new forms of discourse on their way. Early modern philosophy *as such* clearly displays this basic desire. In those days, another word for 'library' was 'metaphysics' (as in the days of Socrates another word for 'textual knowledge' was 'sophistry'). David Hume (1970) summarized the typically modern desire to do away with the library (that is, with metaphysics) altogether when in *An Inquiry Concerning Human Understanding* he states that, while sparing mathematics and empirical science, one should destroy everything else:

"When we run over libraries, persuaded of these principles, what havoc must we make? If we take in our hand any volume of divinity or school metaphysics, for instance, let us ask, *Does it contain any abstract reasoning concerning quantity or number?* No. *Does it contain any experimental reasoning concerning matter of fact and existence?* No. Commit it then to the flames, for it can contain nothing but sophistry and illusion".

This rather extreme attitude can be understood as a revolt against all kinds of non-empirical weights and burdens that are perceived as standing in the way of true intellectual advancement. On other occasions, libraries are being destroyed (or at least threatened with destruction) unintentionally. During the first half of the Twentieth Century, for example, the German language constituted the international scholarly *lingua franca* (French and English coming second). After World War II, however, scholarly discourse migrated into English. And now, as fluency in German and French is declining, enormous archives, written in German and French, are seriously at risk – unless they are 'saved' by translations into English. What is our responsibility, as scholars and scientists, towards the archives of science? Should we save them by converting them to updated formats?

The emergence of electronic archives adds a new dimension to this dilemma. The pace in which information migrates into new formats is increasing. Indeed, languages such as German or English could be regarded as a 'format', but now we are faced with updating and outdating of *electronic* formats, both in terms of hardware (technology) and in terms of software (applications). Documents will have to migrate continuously to updated formats in order to escape the fate of becoming irretrievable or simply unreadable. Are we responsible for the survival of scholarly information stored in obsolete formats? Can we afford to witness these devastating 'digital fires' without doing anything? Is it something we should deplore or something we may rather experience with a sense of relief? Foucault's idea that the opening up and analysis of libraries is a form of 'archaeology' is now truer than ever – as each library becomes a collection of materials in many different mediums and formats (most of them 'outdated').

Ethical styles in conflict

Authors may not only perform different functions, many of them will also practice different genres. For example, they may write and publish for colleagues (experts), but for broader audiences as well, as was already mentioned above. In the first case, they are likely to produce texts that are short, while a considerable portion of it will consist of numbers, technical terms and mathematical formulae. In the latter case, they will prefer ordinary, colloquial, readable, perhaps even elegant prose. A good example of a scientific writer who was very successful in both genres, was James Watson, who in 1953 (together with Francis Crick) published the famous article on the structure of DNA. It could only be read and appreciated by a relatively small number of experts. In 1968, however, Watson published his bestseller *The double helix*. In this book a much broader audience is addressed. It typifies science as a fascinating endeavour, and the scientist as a challenging role model, counterbalancing popular images of dull lab work (Van Dijck 1998). Moreover, in this book a new ethical style is introduced and justified. Science is defined in terms of competitiveness, rivalry and ambition, rather than altruism and selflessness. In this respect, the kind of ethics that was conveyed by the quote taken from *Jurassic Park* at the beginning of my paper was already rather old-fashioned. Watson's book "purportedly promoted a new professional ethic which encourages competition and justifies any means to attain a set scientific goal" (Van Dijck 1998, p. 40). Its function was to display an example of a new scientific style, which even involved ruthless predation on other people's work, minimal courtesy to supporting colleagues and peers, defiance of troublesome data and a positive contempt for traditional intellectual concerns. Without a sign of bad conscience, Watson relates for example how he secretly cajoled people into giving him information. He simply changed the rules of the game. Rather than obeying the old rules of courtesy, collegiality and openness (or even intellectual communism, as Robert Merton called it), he promoted new methods such as competition and secrecy. His major victim was Rosalind Franklin, whose decisive contribution to the discovery of the structure of DNA was until recently grossly undervalued. Once again, the conflict over ethical style eventually focused on issues concerning scholarly communication such as: sharing or concealing unpublished results, publishing, citing and acknowledging scholarly work. Thus, the question is not: 'publish or perish', but rather 'publish strategically or perish", at the right moment and in the right kind of journal, for example. Timing and communicative strategies are important. Watson unequivocally advocated the idea of science as a race, an ethic that is still visible in '(the press covering of) the important research initiatives of the present, such as the Human Genome Project. In 2000, for example, Celera's president Craig Venter (already mentioned above) proudly announced that his company had beaten its rival, the public Human Genome Project in a long, closely watched race. Yet, the shift in scientific ethos that Watson advocated and that can indeed be recognized in the competition-driven academic community of today, does not invalidate the traditional goal of science, namely the quest for a better understanding of the world. Competitive motivations can be perfectly valid ways to attain cognitive goals.

Is there a (technical) solution to every (philosophical) problem?

Instead of simply being overwhelmed by the complexities and perplexities of contemporary scholarly discourse, professionals, notably journal editors, have tried to formulate policies that might allow us to deal with the issues at hand. A new – interdisciplinary – field has emerged: publication ethics. The professional scholars involved try to formulate principles of conduct regarding issues such as sponsorship (commercialization), authorship (notably: multiple authorship) and accountability. A group of editors of medical journals met informally in Vancouver, British Columbia, in 1978 to establish guidelines for the format of manuscripts submitted to their journals. The group became known as the *Vancouver Group*. It expanded and evolved into the *International Committee of Medical Journal Editors* (2001). Among the issues to be considered they listed for example the problem of redundant or duplicate publications. Their guideline stipulates that this practice is admissible only if the editors of both journals have approved it and the paper for secondary publication is intended for a different set of readers.

The focus, however, is on authorship, perhaps the most sensitive and complicated topic in contemporary research ethics. According to their guidelines, all persons designated as authors should qualify for authorship, and all those who qualify should be listed. Each author should have participated sufficiently in the work to take public responsibility for appropriate portions of the content. Authorship credit should be based on substantial contributions to conception and design, or acquisition of data, and should involve final approval of the version to be published. Acquisition of funding, the collection of data or general supervision of the research group does not justify authorship. Authors should provide a description of what each contributed. All others who contributed to the work should be named in acknowledgments. Members of research groups involved in multicentre trials should either meet the criteria as stipulated, or simply be mentioned in the acknowledgments. The order of authorship should be a joint decision of the co-authors.

This has certainly been an important initiative, but I would say that its main importance lies in *listing* and *defining* the problems and issues involved, rather than in solving them. For example, it may in practice be quite difficult to tell exactly what a 'substantial contribution' means and where the difference lies between 'supervising' a research group and 'contributing' to the conception and design of the research group's research. The exact meaning of these terms is bound to be context-dependent. The way these criteria are used is likely to differ from discipline to discipline, or even from research group to research group. Indeed, for several years now, I have been teaching publication ethics to groups of PhD students coming from different disciplines. These groups constitute samples of researchers that are active within one particular university setting, the University of Nijmegen, The Netherlands (KUN). But although most, if not all, of the research groups represented in this sample will tend to agree with the principles as formulated by the Vancouver Group, the range of different interpretations and conventions as described by the PhD students enrolled in my course is simply astonishing. To a certain extent, these differences will reflect power relationships and generational conflicts within particular research areas, and in that sense they may be highly dependent on the personalities and practices of individual scientists involved. But as a rule, these differences tend to reflect more fundamental (and therefore more interesting) differences in the ways these branches of research are actually organized. Indeed, in this area as well as in others, moral conventions and interpretations are often technology-induced. Areas of research using complicated and expensive equipment, for example, and relying on broad collaborative networks and team-based consortia with members in many different roles, will often have a less restricted understanding of what authorship means in comparison to other disciplines (which means that they will be less hesitant in listing someone as 'author'). In disciplines where the dependency on technology is relatively limited (such as in philosophy), however, one-author articles are far from outdated, on the contrary: they still constitute the 'normal' case.

The Vancouver guidelines will certainly help us to clarify some of the problems involved in authorship. Establishing and disseminating good professional practices in this area is important. Yet, scholarly publishing as such has dynamics of its own, and the uncertainties involved in scholarly publishing tend be reproduced, rather than 'solved', by the terms and principles stipulated by these and similar guidelines. It would be highly naive to claim that we know what an author 'is' simply by stating that he or she should have contributed 'substantially'. The Vancouver guidelines allow for a considerable amount of variation. The difference, for example, between 'earned authorship' and 'honorary authorship' is a gradual or scalar difference (Macrina 1995, p. 77). Different disciplines and fields of research will develop their own conventions within the general Vancouver framework. And there are even broader, cultural differences involved in a concept like authorship. Whereas the 'continental' idea of authorship tends to focus on *intellectual* authorship, for example, in the Anglo-Saxon realm the focus is rather on authorship as a financial and economical category (copyright).

Professional ethics and the virtue of self-denial

The finishing line of scientists regarding their work as a kind of race is the time of publication in an acknowledged, international, peer-reviewed journal. What do scientists hope to achieve by publishing their work? The answer is that they strive for recognition. The best way to acknowledge the achievements of scientific authors is by citing them. The ultimate reward, for scientists, is perhaps to occur (as often as possible) in the citation index. "I was cited, therefore I exist". To cite the work of those whose concepts, methods and ideas we used is not merely a form of courtesy, but rather a form of 'fairness'. But can science really be fair? Can we, for example, be fair to our competitors? Robert Merton, the famous sociologist of science, was emphasizing the lack of fairness in citation practices when he described what he referred to as the Matthew effect in science:

"For unto everyone that hath shall be given, and he shall have in abundance; but from him that hath not shall be taken away even what he hath" (Mt 13:12).

Most articles published will be cited just a few times, and then they will be forgotten completely. Most authors will be read and cited by only a limited number of readers. Some articles, however, will be cited more often and the number of citations may even reach a critical limit. Beyond that limit, the number of citations is bound to increase dramatically. They will receive hundreds or even thousands of citations. Colleagues will continue to cite them for twenty or thirty years, until the paradigm the author helped to create becomes extinct. An author may publish an article that really makes his name, although the time and effort spent on writing it may not significantly exceed the amount of time and effort spent on publications that are treated less respectfully. Eponymy (mentioned above) is one of the causes of this phenomenon, but also priority ('publish timely') and visibility ('publish strategically'). But even eponymy is a rather technical function of authorship. It is simply a convenient way to refer to a test, an illness or a bodily part. Most authors will remain relatively anonymous and even the identity of authors that are cited quite often will disappear behind their name. Hardly anyone who refers to the Stroop effect, for example, will know anything about the individual bearing the surname Stroop. Thus, as a rule, the desire for recognition will remain unsatisfied. Is there something we can do about this? Should we try to make science fairer?

I think it is a mistake to regard science as an endeavour that has to be tailored and constrained by means of ethical rules and norms external to science as such. Rather, science should be looked upon as an intrinsically moral phenomenon. Not only because scientists often want to contribute to the progress of mankind and the quality of human life, but also because doing science and being educated as a scientist constitutes a kind of *Bildung* in itself. According to Friedrich Nietzsche, science is a powerful tool developed by man to educate and discipline himself. Scientific training is basically a training in self-control. Virtues involved in practicing a science, such as unprejudiced open-mindedness, patience, precision and reliability, are moral values. Science fosters an attitude of self-criticism and perseverance. According to Nietzsche, a true scientist is someone who is really willing to adopt a self-critical, unprejudiced stance, someone who will not allow himself to be dominated by stereotypical views, narcissism and prejudice, but is always willing to put his theories to the test and to enter into an open debate with others, someone who is susceptible to criticism. According to Nietzsche, the true scientist has but one desire: not to deceive, neither himself nor others. That is, science is an intrinsically moral phenomenon. And perhaps the most important scientific virtue of all is self-denial. *"Was liegt an mir!"* It is not me that counts! According to Nietzsche, this phrase more or less sums up the ethic of being 'in science' (Nietzsche 1980, § 547). And a similar phrase occurs in the work of Michel Foucault: "Qu'importe qui parle?" Science is, first and foremost, simply discourse: a conversation or intellectual exchange that already existed before I decided to participate in it and that will continue to exist when I am no longer present or able to contribute to it. *On parle... Man spricht...* Most, if not all, of the words, terms, phrases and arguments a scientific author uses, have been invented by others. In most cases, scientific authorship comes very close to anonymity. There is a certain moral quality in the stoical acceptance of this fact.

Acknowledgements

I want to express my gratitude to my colleague Luca Consoli for his critical reading and his useful suggestions.

References

Crichton, M., 1991. *Jurassic Park*. Arrow Books, London.
Foucault, M., 1995. Q'est-ce qu'un auteur. *In: Dits et écrits. 1. 1954-1969*. Gallimard, Parijs, 789-809.
Fraser, G., 2000. *Antimatter: the ultimate mirror*. Cambridge University Press, Cambridge.
Hawking, S.W., 1988. *A brief history of time: from the big bang to black holes*. Bantam, Toronto.
Hume, D., 1970. *Enquiries concerning the human understanding and concerning the principles of morals: reprinted from the posthumous edtion of 1777*. Clarendon Press, Oxford.
International Committee of Medical Journal Editors, 2001. *Uniform requirements for manuscripts submitted to biomedical journals: writing and editing for biomedical publication*. ICMJE Secretariat office, American College of Physicians, Philadelphia. [http://www.icmje.org/]
Macrina, F.L. (ed.) 1995. *Scientific integrity: an introductory text with cases*. ASM Press, Washington DC.
Merton, R.K. and Storer, N.W., 1973. *The sociology of science: theoretical and empirical investigations*. University of Chicago Press, Chicago.
Nietzsche, F., 1980. *Morgenrote: Sammtliche Werke, Kritische Studienausgabe*. DTV, Munich.

Todes, D.P., 2002. *Pavlov's physiology factory: experiment, interpretation, laboratory enterprise*. John Hopkins University Press, Baltimore.
Van Dijck, J., 1998. *Imagenation: popular images of genetics*. New York University Press, New York.
Venter, J.C., Adams, M.D., Myers, E.W., et al., 2001. The sequence of the human genome. *Science,* 291 (5507), 1304-1351.
Vorländer, K., 1977. *Immanuel Kant: der Mann und das Werk*. 2nd edn. Meiner, Hamburg.
Watson, J.D., 1968. *The double helix: a personal account of the discovery of the structure of DNA*. Atheneum, New York.
Watson, J.D. and Crick, F.H.C., 1953. Molecular structure of nucleic acids: a structure for deoxyribose nucleic acid. *Nature,* 171 (4356), 737-738.
Wippern, J. (hrsg.), 1972. *Das Problem der ungeschriebenen Lehre Platons: Beiträge zum Verständnis der platonischen Prinzipienphilosophie*. Wissenschaftliche Buchgesellschaft, Darmstadt.
Zeller, E., 1980. *Outlines of the history of Greek philosophy*. 13th edn. Dover Publications, New York.
Zwart, H., 2001. *De wetenschapper als auteur: geschiedenis en toekomst van het wetenschappelijk communiceren*. SUN, Nijmegen.

---

[1] A word of caution is due in order to avoid confusion: the word 'science' with its modern connotations may not seem fully appropriate when dealing with the origins of Western philosophy. Since, however, that endeavour can confidently be seen as the beginning of a process of rational inquiry that has led to our modern concept of science, I have chosen to stress the continuity between ancient and modern 'science'

[2] An illustrious exception can be found in Leonardo da Vinci. Although he is regarded as a typical 'homo universalis', and he was acquainted with the whole corpus of the knowledge of his time, he also was an observer and made considerable original contributions in the empirical sense. We can see him as a transition figure.

5b

Comments on Zwart: Professional ethics and scholarly communication

Tjard de Cock Buning[#]

Introduction

I appreciate professor Zwart as a brilliant scholar who seeks his intellectual challenges in overarching philosophical perspectives. However, when I read his paper I realized that our approaches to the topic of professional ethics are completely different. I expected at least three focus points in the tradition of the Reflective Equilibrium by Rawls (1971) and others in The Netherlands like Heeger (Van Willigenburg and Heeger 1989) and De Cock Buning (1998): (i) a spectrum of representative descriptions of boundary cases regarding professional communication by scholars, (ii) some moral intuitions, and (iii) finally a set of relevant ethical principles to bring the boundary cases to an ethical judgment. What is presented, however, are four unrelated parts, the former two dealing with intuitions and descriptions of circumstances and cases, and the latter two addressing ethical aspects.

Jurassic park

The author of Jurassic Park is cited when Zwart describes the historical change in academic freedom and communication. According to Zwart this description is false. If he is right, I wonder whether we should take serious notice of this apparent false description. The argument, however, does not come back in the rest of the text.

A case is described of a researcher who was pressured by a company that financed her research project. No indication is given for options in relation to a professional ethical frame of reference, nor in the text that follows. The case is left with an open end. Why?

The more favourable approach, in my opinion, is the one developed in the Science and Society programme of the Royal Academy of Science. From 100 codes of conduct Bout and De Cock Buning (1998) deduced a shared structure. This generalized frame became the leading concept in the code of conduct for biologists, medical practitioners and psychologists and for the Project on codes of conduct at Wageningen UR. We learned that the professional codes of conduct and their huge variety of more or less particular provisions, amount to four basic principles: *Integrity*, *Competence*, *Respect* and *Responsibility*. Therefore, when we return to the case of the researcher put under pressure by the funding company, one could 'solve' this case by analysing the problem as a consequence of improper implementation of the 'responsibility' structures. The university, i.e. the legal office of the university, should be the responsible party in deals with the industry. However, if one believes

---

[#] Department of Biology and Society, Free University, De Boelelaan 1087, 1081 HV Amsterdam, The Netherlands. E-mail: Tjard.de.Cock.Buning@falw.vu.nl

that the individual researcher must negotiate with the industry, one will end up in the uncomfortable situation of the described case.

Communication in a changing ICT world

Fundamental and challenging questions are raised by Zwart, such as: "What is the relevant definition of an author?" and "What is the relevant definition of a publication?" However, no suggestions regarding a possible answer are given.

Regarding the question of 'true' authorship, many attempts have been made in professional codes of conduct (e.g., by the International Committee of Medical Journal Editors 1997). Most of them state that an author must meet several criteria. He or she must provide a substantial contribution towards (1) the idea and design of the project, (2) the analysis and interpretation of the data, (3) the drafting and reviewing of the intellectual content and (4) the approval of the final version for publication. The first three are operational aspects of the competence principle and the last one is related to the demand for responsibility and accountability.

Ethical styles in conflict

At this point, I expected a broad spectrum of relevant ethical notions that would give us various options to deal with the uncomfortable cases. Instead of such an analysis we received again a description of a new set of ethical notions, viz. the set of criteria used by Watson: competition, rivalry and ambition (as opposed to Merton's altruism, solidarity and selflessness). Why doesn't Zwart position these two different styles of ethics in ethical frames? These two styles may be entered in the scheme developed by Zweers (1995; 2000), who distinguishes five worldviews (see Table 1). Watson clearly fits into the 'ruler' position (aggressive language, technocratic optimism) and Merton takes the middle position (responsible stewardship, win-win pragmatism). From this conceptual perspective, one could very well position Watson and his criteria under the ruler position on the one hand, and at the same time position the agricultural practice (i.e. criteria of our host, Wageningen University) under the stewardship position on the other hand. Such an approach enables us to analyse better the distinctions and equalities within the same conceptual framework. Although still descriptive it would be descriptive in relation to an ethical/societal framework.

Table 1. Five types of worldviews, freely interpreted after Zweers (1995; 2000)

Types of worldview	Characteristics		
Commander	To control	Man against nature	Technocratic
Enlightened ruler	To manage	To improve	Accountable
Steward	In service of	Conservation	Win-win
Partner	To develop	Man and nature	Equality
Ecocentrist	To take part in	Man is nature	Empathy

Personally I doubt whether the Watson example is the relevant example. Watson explicitly chose to write a bestseller. He knows that the public likes to read about hidden conflicts. So he highlights these conflicts. His book is written as if it were a crime story. Actually, all professional codes of conduct agree with Merton's approach and not with Watson's perspective. The ethos that Watson seems to advocate serves, on the contrary, as the one to be condemned. Again, I wonder which message Zwart wants to convey. Is he describing the anti-ethos? But does he advocate it? There is an intriguing absence of argument.

The virtue of self-denial

Zwart suggests that scientists aim for recognition and the satisfaction of recognition. He considers the Matthew effect unfair. He directs our attention to the eponymy. The validation of this analysis is not stated in empirical evidence, but is probably (at least) not contrary to his own experience. I doubt, however, the general validity of his analysis.

When I take my personal career as situational evidence, I come to quite a different conclusion. I received the honour of a PhD *summa cum laude* on a biological topic that only ten people in the world would appreciate. They did; and I was, and still am, satisfied. In line with this argument, I claim that there is no straightforward relation between the fame of the Nobel laureates in the domain of natural sciences and the number of people who understand their articles. In other words, one should be careful not to confuse satisfaction coming from professional colleagues and satisfaction coming from fame attributed to the scientist by media machinery. The last type of satisfaction depends upon one's character. Some prefer to become a public figure. I would like to consider this latter interpretation as a part of psychology and not as a part of professional ethics.

At the end Zwart cites Nietzsche to sum up some of the basic values of the individual scientist. As may be noticed, they all correspond well with the above-mentioned four principles that lie at the basis of professional codes of conduct. Although he clearly sympathizes with these statements of Nietzsche, he does not take the consequence to criticize Watson from this position, nor does he analyse the problem of justifiable authorship in this context, nor the various modes of publications in ICT. On the contrary, he comes up with a poetic statement that scientists are close to anonymity, which is only slightly true when one confuses public fame with recognition by one's colleagues.

Commenting upon a text is an easy job. Presenting an alternative is another story. I would like to take the opportunity to propose my answers to the cases presented by Zwart. I would place the cases in the context of the conceptual framework of 'professional ethics'. I would ask myself the question whether in a specific case the basic notions of competence, integrity, respect and responsibility are of any relevance to the actors in order to organize their behaviour professionally. From this perspective I can easily analyse the case of professional authorship according to the principles of *competence* (in order to grant an authorship one should be competent as to the manuscript, i.e. directly related to the research), *integrity* (no false games and hidden rewards should intervene in the list of authors) and *responsibility* (all authors should take full responsibility for the presented manuscript).

The case of internet publications can be analysed by *competence* (some professional provisions should be made to guarantee the competence of the authors) and *responsibility* (the internet publisher must somehow organize his responsibility for the quality of the publications).

The professional approach against contractual pressure is guided by the principles of *responsibility* (the responsible actor is not the researcher but the legal representative of the research institute) and *integrity* (everyone should guard his/her own role).

The topic of scientific satisfaction comes simply down to a proper acknowledgement of mutual *respect*. Providing options to guarantee that persons who deserve respect will meet respect should be obvious in a professional research setting.

References

Bout, H.J. and De Cock Buning, T., 1998. *Leidraad voor het opstellen van een adviserende beroepscode.* Leiden University, Leiden. Rapport Proefdier & Wetenschap no. 10.

De Cock Buning, T., 1998. Moral costs of animal experiments. *In:* O'Donoghue, P.N. ed. *The ethics of animal experimentation.* European Biomedical Research Association, London, 45-50.

International Committee of Medical Journal Editors, 1997. Uniform requirements for manuscripts submitted to biomedical journals. *JAMA: the Journal of the American Medical Association,* 277 (11), 927-934.

Rawls, J., 1971. *A theory of justice.* Belknap Press, Cambridge.

Van Willigenburg, T. and Heeger, F.R., 1989. *Justification of moral judgement: a network model.* Societas Ethica, Hanover. Societas Ethica Jahresbericht.

Zweers, W., 1995. *Participeren aan de natuur: ontwerp voor een ecologisering van het wereldbeeld.* Van Arkel, Utrecht.

Zweers, W., 2000. *Participating with nature: outline for an ecologization of our world view.* International Books, London.

6a

Some recent challenges to openness and freedom in scientific publication

David B. Resnik[#]

"The right to search for truth implies also a duty; one must not conceal any part of what one has recognized to be true."
Albert Einstein, engraved on his memorial statue at the National Academy of Sciences, Washington, DC.

Introduction

Most scientists probably share Einstein's commitment to searching for and revealing the truth. This commitment implies a variety of ethical norms and values, including honesty, integrity, objectivity, openness, freedom, carefulness and fairness (Shamoo and Resnik 2002). However, professional ambitions and rivalries, financial interests, intellectual property disputes, ideological agendas and other social, economic and political influences can disrupt or derail the quest for the truth (Ziman 2002; Kitcher 2001). Since modern research is a social phenomenon, it is not possible to eliminate these social, economic and political factors from the scientific milieu (Kuhn 1970; Hull 1988; Longino 1990). Even so, scientists, research sponsors and academic institutions should strive to maintain a strong commitment to the search for the truth, and they should develop policies and institutions that minimize the impact of external biases and influences on research (Shamoo and Resnik 2002).

Many of the important ethical problems and issues in scientific research reflect the clash between science's ethical ideals and these non-scientific (external) influences in the contemporary research environment (Resnik 1998). Nowhere has the clash between scientific and non-scientific values been more evident than in the area of publication and the dissemination of information, where private and government interests may conflict the scientific commitment to search for and reveal the truth. This paper will discuss several recent problems for openness and freedom in scientific publication related to the private sponsorship of research and the threat of bioterrorism. The paper will also suggest some potential solutions to these problems.

Private industry and the suppression of research

Private industry sponsors more than half of all research and development (R&D) conducted in the world. In the year 2000, private industry accounted for roughly 60% of the $200 billion that the United States (US) spent on R&D (Shamoo and Resnik 2002). Private investment in R&D, which had been less than the public investment in R&D throughout the 1960s and 1970s, rose significantly in the 1980s and 1990s,

---

[#] Brody School of Medicine, East Carolina University, 2S-17 Brody Building, Greenville, N.C., 27858, USA. E-mail: resnikd@mail.ecu.edu

while public investment rose only marginally. Most of the increase in private R&D was due to the continued growth of the pharmaceutical industry and the emergence of the computer and biotechnology industries. Since businesses have strong economic motives to invest in R&D and government budgets constrain increases in public investment in R&D, it is likely that the private sector will outspend the public sector for many years to come.

The infusion of private money into science has benefited researchers, businesses, universities and society, but it has also taken a toll on openness and freedom in research. Businesses aim to make a profit and to produce goods and services. They do not search for the truth for its own sake; they regard R&D as a necessary means of achieving financial and practical goals, and they are more than willing to restrict openness or freedom in order advance their primary goals.

If a scientist is employed by a private company to conduct research, the company will usually make him sign a contract in which he agrees that the company owns all of his data and has the authority to review and approve any of his publications. Even a scientist employed by a University may sign a contract with a private company that sponsors his research, which gives the company the right to review his research and approve of any publications. Although some universities do not allow their employees to sign these contracts, many do. A scientist who violates the provision of one these contracts can face adverse legal consequences, including civil liability for breach of contract or negligence as well as criminal liability for disclosing trade secrets.

There are at least three reasons why a private company would want to prevent a scientist from publishing research sponsored by the company. First, the company might seek to block publication in order to protect its intellectual property rights. Publication of information used to develop a patentable invention may count as a prior disclosure. Patent laws in the US and Europe require that the invention be novel. If the invention has already been publicly disclosed, then it will not meet the novelty requirement and it will not be patentable (Miller and Davis 2000). Although patenting can delay publication of scientific and technical information, in the long run it encourages public disclosure because the patent application becomes a part of the public record. Thus, although intellectual property interests can pose a short-term threat to the search for the truth, in the long run they benefit science by providing incentives for inventors, investors and entrepreneurs.

A second reason why a company might want to block research is that it does not want its competitors to discover its new products, business strategies or other trade secrets. It wants to maintain secrecy in order to secure a competitive advantage. Although the desire for trade secrecy can inhibit the search for the truth in the short term, in the long run many of the secrets that a company keeps will become public knowledge as the company places goods and services on the market, implements business strategies and discloses its secrets. Other secrets may be discovered by legal means, such as reverse engineering or independent research (Shamoo and Resnik 2002). A more troubling reason why a company might want to block publication is that publication of adverse data or results may undermine its ability to market a particular good or service. For example, if a company sponsors research that compares its drug to competing drugs, and the research demonstrates that its drug is no better than the competitors', it might try to suppress publication of the research. Or even worse, if a company sponsors a study that shows that its product is dangerous, it might try to suppress this research as well. Three highly publicized cases have illustrated problems with the suppression of research.

In 1994, the US Congress held hearings on the tobacco industry. A Congressional committee subpoenaed the testimony of Drs. Victor DeNobel and Paul Mele, who conducted research for Philip Morris on nicotine addiction in the 1980s. DeNobel and Mele testified that their research proved that nicotine is highly addictive and that they discovered substances that increase the addictive properties of cigarettes, while reducing the adverse cardiovascular effects of cigarettes. The purpose of their research was to develop a substitute for nicotine that would make cigarettes more addictive. DeNobel and Mele, who were employees of Philip Morris, were not allowed to discuss their work with other employees or colleagues. Animals used in their research were brought into the laboratory under covers. The two scientists tried to publish the results of their work in *Psychopharmacology*, but Philip Morris learned about the paper and forced DeNobel and Mele to withdraw their paper. The company also dismissed the two scientists and shut down their laboratory. DeNobel and Mele had signed a contract with Philip Morris in which they agreed never to discuss their research without the company's permission, but Congressman Henry Waxman arranged for the two scientists to be released from this agreement so they could testify before Congress (Hilts 1994).

In 1995, the Boots Company made Dr. Betty Dong withdraw a paper on drugs used to treat hypothyroidism, which had been accepted by the Journal of the American Medical Association. Boots had funded Dong's research, which compared its drug, Synthroid, to some generic drugs. Boots found that Synthroid was not safer or more effective than the generic drugs and that the US could save millions of dollars a year if patients switched from Synthroid to one of the generic drugs. Dong, who was a clinical pharmacologist at the University of California at San Francisco, had signed a contract with Boots giving the company permission to review her results and prevent her from publishing her work, without written permission. The company threatened to sue Dong and also spent two years attempting to discredit her research. To avoid a lawsuit, Dong withdrew the paper. However, the company eventually relented, and two years laters Dong published her results in the *New England Journal of Medicine* (Wadman 1996; Shamoo and Resnik 2002).

From 1993 to 1995, Dr. Nancy Olivieri and her colleagues at the University of Toronto and Toronto General Hospital conducted research on a drug used to treat thalassaemia, deferiprone. Their research was sponsored by Apotex Inc., a Canadian pharmaceutical company. In 1995, Olivieri and her collaborators published an article on deferiprone in the *New England Journal of Medicine*. The study reported that the drug was effective at reducing total body iron stores in thalassaemia patients and had manageable side effects. A few months after they reported these positive findings, they observed that liver iron stores in many of their patients were reaching dangerous levels, which could lead to heart failure or death. Olivieri wanted to notify the hospital's Research Ethics Board (REB) about this problem, so that the consent forms could be revised and patients could learn about this new risk. Apotex tried to prevent Dr. Olivieri from reporting her concerns to the REB. She did eventually notify the REB, but after she did, the company terminated the study and withdrew all the supplies of the drug from the hospital pharmacy. The company also threatened to bring litigation against Olivieri if she would decide to tell patients, regulatory agencies or the scientific community about her concerns. Several other studies confirmed Olivieri's concerns about the drug. She continued to receive letters from the company threatening legal action, and she withdrew some presentations on the drug she had planned to make at scientific meetings (Olivieri 2003).

Chapter 6a

In 1998, Apotex, which was negotiating a large donation to the University and the Hospital, pressured these institutions to take actions against Olivieri. Olivieri had assumed that the Hospital and the University would take her side in the dispute with Apotex, but these two institutions denied there was a problem, sought to delay public awareness of the problem, tried to divide Olivieri from her colleagues, sought to discredit her work, and even tried to have her dismissed from her position. Finally, in January 1999, an international group of ethicists and scientists lent their support to Olivieri and prevented her from being dismissed. Olivieri reached an agreement with the Hospital and the University, clearing her of all allegations. In 2000-2001, a commission from the Canadian Association of University Teachers investigated the incident (Olivieri 2003).

Clearly, the suppression of research by private industry represents a significant threat to the search for the truth, since a private company could use this strategy to the published research record, to keep undesirable results a secret or to control the conduct of scientists. How should the scientific community respond to this problem? Private companies should be encouraged to sponsor R & D, provided that they adhere to some rules for publication. To develop these rules, we should distinguish between different types of research: industry-sponsored research conducted in a university (or academic) setting and industry-sponsored research conducted in private laboratories. Since openness and freedom are vital to the academic environment and university-based research, universities should not allow their faculty to sign contracts that grant private companies the right to block the publication of research conducted on campus. No private company should be able to suppress academic research. All contracts signed by academic researchers with private companies should give them the right to publish data as soon as it is necessary to promote the advancement of research or address important public-health or safety concerns. Additionally, academic institutions should support researchers who become involved in disputes with private companies about publishing data and results (Nathan and Weatherall 2002).

What about research conducted in private laboratories? Should governments enact laws that forbid private companies from signing employees to contracts granting the company the right to suppress publication of research conducted for the company?

Although it would be desirable to encourage private companies to guarantee the same degree of freedom and openness that one finds (or expects to find) in academia, restrictions on the contracts that private companies sign with their employees would be unwise. First, in the US (and possibly in other countries), such restrictions would run into legal challenges. In the US, laws that restrict the freedom of private contracts must have a reasonable relationship to the public interest, and they should be neither arbitrary nor discriminatory (Nebbia v. New York 1934). Private companies that sponsor research could argue that laws that restrict the contracts they sign with scientists would not serve the public interest because they would discourage companies from conducting research. Second, one might argue that a private laboratory is not the same as an academic institution, because private corporations, unlike academic institutions, are established in order to make a profit. Although academic institutions thrive on freedom and openness, control and secrecy are essential to private businesses. Businesses need to maintain trade secrets to build and maintain competitive advantages. Since trade secrecy should still protect private research conducted in private laboratories, it would be unwise to require businesses to sign researchers to contracts that could undermine trade secrecy. Thus, the research the DeNobel and Mele did for Philip Morris is fundamentally different from the research that Dong and Olivieri conducted for pharmaceutical companies. In both instances, private money supported the research, but in the second instance, the research was conducted in an academic environment. Only in this second instance should there be limitations on private contracts with researchers that are designed to promote freedom and openness.

Private industry and access to data from published research

Another important issue where private industry poses a threat to the search for the truth concerns access to data after publication. Several surveys have shown that many scientists working in academic and non-academic setting frequently refuse to share data prior to publication (Blumenthal et al. 1997; Campbell et al. 2002). The main reasons mentioned by respondents to the survey are that they withheld data in order to protect unpublished work, to secure priority, to protect intellectual property rights, or because data-sharing was inconvenient or expensive. These are all good reasons to guard data prior to publication. First, prior to publication, data may be inconclusive or unconfirmed. Premature publication of data or results can have disastrous effects, as illustrated by the cold-fusion controversy (Shamoo and Resnik 2002). In this case, Drs. Stanley Pons and Martin Fleischmann believed that they had discovered a way to produce fusion at room temperature in an electrolytic solution. Pons and Fleischmann presented incomplete descriptions of their methods as results at a press conference before submitting their work for peer review. Scientists around the world scrambled to try to replicate their results, but they were unsuccessful. Many scientists were upset that Pons and Fleischmann published their data through a press conference prior to peer review; other accused them of fraud, negligence or self-deception. The entire episode had a negative impact on the public's perception of science.

Second, important claims to intellectual priority may be at stake when scientists are asked to share unpublished data. Priority disputes have occurred in science for hundreds of years. In science, the credit goes to the researchers who publish first (Merton and Storer 1973). If a researcher shares his or her data with someone else before it is published, he or she may not receive credit for making an important discovery, and the person that received the data might steal the researcher's work and take credit for the discovery. Priority is also important in establishing patent rights. To be patentable, an invention must be novel, non-obvious and useful. An invention that has been previously disclosed through prior publication or use will not be considered novel (Miller and Davis 2000). A researcher who shares or publishes data before filing a patent application may lose his or her legal right to patent the invention. Furthermore, someone else could use shared data to beat the researcher in the race for patent rights. If two researchers both apply for a patent on the same invention, patent offices will award the patent to the first person to conceive of the inception, provided that they have both exhibited due diligence in prosecuting the patent application and reducing the invention to practice (Miller and Davis 2000).

Although researchers often have good reasons not to share data prior to publication, they should share data after publication, if sharing data will not violate the privacy of research subjects. Once a researcher has published the results of his scientific work and received proper credit for his or her accomplishments, he or she should make his or her data available to other researchers. It is important to share data after publication because other researchers may need the data to verify the results, repeat the experiments, to learn how the research was conducted or to stimulate new discoveries and findings (Shamoo and Resnik 2002). In the US, government agencies that sponsor research, such as the National Institutes of Health (NIH), require researchers who receive contracts or grants to share data once the main results of a research project have been accepted for publication (National Institutes of Health 2003). However, these data-sharing rules do not apply to private corporations, which may decide to publish the results of a research project and then charge a fee for access to the data. Even though many scientists share data after publication, some still refuse to share data after publication. One reason why researchers may not share data supporting published results is that one may still make important discoveries by analysing the unpublished data, and some scientists do not want researchers who have not invested time, money and effort in gathering data to take undeserved credit for publications based on the data (Barinaga 2003). Another reason for not sharing unpublished data is that an individual or a group with a political agenda, such as an animal-rights group, could use the unpublished data to harass the researcher.

In February 2001, the public consortium, led by the National Human Genome Research Institute (NHGRI), and Celera Genomics, a private company, published versions of the human genome in the journals *Nature* and *Science*, respectively. The NHGRI deposited its data relating to the human genome in the Genbank, an enormous electronic database that researchers can access for free. Celera, however, refused to deposit its data in the Genbank. Under the terms and conditions negotiated between *Science* and Celera, non-profit researchers were allowed to download data from Celera's website, provided that they agreed not to commercialize or distribute the data. Researchers who planned to use the data for commercial purposes were required to negotiate an agreement with Celera (Marshall 2001). *Science* reached a similar agreement with Syngenta when it published a draft sequence of the rice genome (Marshall 2003).

Many scientists were angry that *Science* decided to publish Celera's paper describing the human genome without requiring the company to make its data freely available. In 2003, a group of leading researchers from the biosciences issued a report on sharing data in the life sciences (National Research Council NRC 2003). The report prescribes rules for sharing data and materials known by the acronym UPSIDE (Universal Principal of Sharing Integral Data Expeditiously). The UPSIDE rules recommend that all scientists who publish research on genome sequences should immediately deposit their entire data set in a public database, such as Genbank. The report also declares that scientists should also share materials pertinent to their research findings and explain how they were obtained. The editors of *Science* and *Nature* have both said that they would abide by the UPSIDE rules. When Celera published its version of the human genome, there were no generally accepted rules for data-sharing following publication. Currently, 45% of journals surveyed do not have a data-sharing policy (Marshall 2003).

Placing restrictions on access to published data also poses a significant threat to openness and free inquiry. Ideally, researchers should make all of their data available as soon as they publish their work. In an ideal world, all data would be freely available to all researchers after publication. But, we do not live in an ideal world. In the real world, someone must pay a great deal of money to produce research data and develop and maintain databases. In the real world, governments cannot afford to fund all research and development (R&D), and researchers must draw on private funding. When private companies invest in R&D, they expect to obtain a reasonable return on the investment. If they cannot expect a reasonable return on their investment, then they will stop sponsoring R&D or they use trade secrecy to protect data and results. Neither of these possibilities bode well for the advancement of science, technology or industry. Over the years, private companies that invest in R&D have used a variety of business strategies to gain a return on their investments, including patenting new inventions, copyrighting original works and selling products and services. In the today's research environment, access to electronic databases plays a crucial role in research and development in biomedicine and biotechnology. As a result of advances in bioinformatics, researchers can use computer programs to search and analyse databases in order to discover patterns and connections (Mauer and Scotchmer 1999; Freno 2001). For example, one can use computer programs to compare a mouse DNA sequence to a human DNA sequence or to determine the relationship between a viral DNA sequence and its protein product. The ability to search and analyse data therefore also has a great deal of economic value in research, since scientists may be willing to pay a considerable sum for access to data and the ability to search and analyse databases. Some private companies, such as Celera, have developed business models for selling information services. When Celera published its version of the human genome, it was planning to charge researchers a fee for the ability to access, search and analyse DNA-sequence data (Marshall 2001).

Let's assume that private companies should be able to obtain a reasonable return on the R&D investments, including their investments in developing and maintaining databases. Given this assumption, we need to ask *how* companies should be able to make money from their investments. Let us also assume that companies should be allowed to make money in the traditional ways, i.e., through patents and copyrights and selling their goods and services. Our question then becomes: should companies be able to make money by selling data services? The key to resolving this issue is to find a fair balance between scientists' interests in access to research data/results and private interest in making money from selling access to data/results. Someone must pay for the initial R&D required to generate the data and the subsequent R&D needed to develop and maintain the database. But who should pay?

For a useful analogy, consider scientific journals. Journals have several options for generating income including drawing income from authors (e.g. pages charges), from users (e.g. subscription fees), from advertising, or from institutional sponsors (e.g. government or private corporations). Most journals draw income from all of the different sources, and very few journals do not charge users a fee for access to articles. Those that do not charge a fee to users usually have a great deal of institutional support. In addition, nearly every journal provides some information for free via public databases of abstracts and keywords. For instance, anyone can search MEDLINE for abstracts of articles on prostate cancer, but to get a copy of the full article, one must pay the journal, copy the article in the library or write the corresponding author. Under this system, there are two tiers of sharing. The first tier offers free access to abstracts, which usually contain information about the significant findings and results. The second tier offers access to full articles, which are usually not free.

A two-tier system for research data might work as follows. The first tier would provide raw data to the public, free of charge. The second tier would provide access to data that have been analysed and embellished. Companies could charge a fee for access to the second tier of data and require users to sign a licensing agreement. The first tier would provide researchers with the information they need to confirm published results, but it would not provide researchers with the extra features that can stimulate new research and innovation. The second tier would have economic value because it would be useful to researchers who want to search and analyse databases. For a relevant analogy, consider legal information, such as judicial opinions from legal cases, legislative statutes and materials, and administrative policies and procedures. Most of these materials are available online for free in the US. However, it is not at all easy to find, search or analyse these free materials. Westlaw, a legal-information company, has developed an immense electronic library of legal resources that is easy to search and analyse. Westlaw charges users a considerable fee for access to its private database, even though most of the materials it has are also available for free (somewhere). I would like to suggest that Westlaw provide private companies with a good model for selling information services: companies can make data freely available but also charge a reasonable fee for access to well-organized, searchable and analysed databases. Of course, for the Westlaw model to work, it is also important that companies develop savvy licensing agreements and that governments provide private databases with adequate protection under copyright law (Freno 2001).

The military and classified research

The problems discussed in the previous two sections of this paper arise as a result of the conflict between the values of openness and freedom and the interests of private industry. Problems can also occur when openness and freedom conflict with national and international security interests. For many years, the US military has sponsored classified research related to national and international security, including research on weapons systems, defence systems, reconnaissance devices, intelligence methods, encryption techniques and military strategies and tactics. The phrase 'loose lips sink ships' aptly describes much of the research sponsored by the US Department of Defense (DOD) or conducted at DOD facilities and laboratories (Dickson 1984). Classified research is conducted under strict secrecy rules and is not published or otherwise shared with the public until it is declassified. For example, in 1994 President Clinton declassified thousands of documents pertaining to secret human radiation experiments conducted and sponsored by the DOD from the late 1940s to the 1980s (Moreno 1999). Although the military has been granted the authority to classify research with implications for national and international security, it does not have the authority to classify basic scientific information not related to national security (Atlas 2002).

Disputes concerning the status of basic scientific information related to cryptography have existed since World War II, when the ability to encode and decode messages proved to be very important in the Allies' victory over Germany and Japan. The science of cryptography has made tremendous advances since World War II as a result of the development of computer encryption and decryption programs. Officials from groups concerned with national security, such as the DOD, the National Security Agency (NSA), the Central Intelligence Agency (CIA) and the Federal Bureau of Investigation (FBI), have tried to control the public dissemination of cryptography research in order to prevent hostile foreign governments, terrorists groups or criminal organizations from having access to advanced encryption technology. On the other hand, scientific organizations with an interest in the sharing of information as well as private companies with an interest in encryption technologies have attempted to keep basic scientific information out in the open.

The US Congress has passed laws that give the US President the authority to restrict the exportation of technologies to foreign governments if they are likely to aid the development of weapons of mass destruction, support international terrorism, increase the possibility of conflict or prejudice arms-control efforts (Ackerman 1998). However, practical and legal obstacles may prevent the President from effectively using these laws to stop the dissemination of encryption information. From a practical point of view, it is almost impossible to stop the exportation of encryption technology, since computer source codes are very easy to transmit electronically. From a legal point of view, a ban on exports of encryption programs could be an unconstitutional interference with the freedom of speech. During the 1990s, Congress considered several bills that would have given the government a way to decrypt all encrypted messages. The idea was to require that all encryption devices or algorithms contain key-recover techniques, which would allow someone with the right information, i.e. a key-recovery agent, to decrypt the encrypted message. Congress also considered several bills that would make key recovery voluntary rather than mandatory. Civil-rights groups, privacy groups, business groups and computer-science organizations opposed key-recovery legislation. The National Research Council raised some issues relating to the security threats posed by key recovery technology. For instance, what would happen if the wrong person got access to a recovery key? So far, the US and Europeans countries have not adopted key-recovery legislation, although they have attempted to control exports of encryption technology (Ackerman 1998).

Bioterrorism is a relatively recent threat to national and international security. Although governments have developed biological and chemical weapons for many years, in the 1990s military and political leaders, biologists, political scientists, public-health experts and security analysts became increasingly concerned about the possible use of biological or chemical weapons by terrorists on civilian populations. The horrors of the large-scale use of mustard gas during World War I led to the adoption of the Geneva Protocol in 1925, which forbids the use of bacteriological and chemical weapons in war. In 1972, dozens of countries signed the 1972 Biological and Toxic Weapons Convention (BTWC), a treaty that prohibits the development or possession of biological weapons. Although many countries, including the US, Russia and China, have signed the BWTC, as many as 17 countries currently possess or are developing biological weapons (Cole 1996). Although Russia claims that it does not have a bio-weapons programme, scientists in the former Soviet Union had an extensive bio-weapons programme that studied the use of anthrax, botulism, the plague, the Ebola virus and the Marburg virus (MacKenzie 1998).

Iraq used chemical weapons during its war with Iran during the 1980s and during its suppression of a Kurdish uprising in 1988. Iraq has acknowledged to United Nations Weapons Inspectors that it had Scud missiles tipped with biological warheads during the 1991 Persian Gulf War (Cole 1996). After that war, weapons inspectors attempted to determine whether Iraq had biological or chemical weapons or was developing them. Iraq denied that it had any of these weapons and kicked out the inspectors in 1998.

Following a series of debates at the United Nations about Iraq's weapons programme, the futility of inspections and the potential use of these weapons by terrorists, the US and the United Kingdom (UK) invaded Iraq in late March of 2003 to enforce UN Security Council Resolution 1441, to find and eliminate these alleged weapons, and to remove Saddam Hussein, the leader of the Iraqi government, from power. (This essay will not engage in a debate about the moral or political justification of this military operation, or lack thereof.) At the time of the writing of this essay, neither the US nor the UK have found any conclusive evidence of biological or chemical weapons in Iraq, but it could take months to conduct a thorough search of the country.

The leaders of the US and UK were concerned that the Iraqi regime might provide chemical, biological or nuclear weapons to terrorist groups, such as Al-Qaeda, the organization which is held responsible for dozens of attacks on civilian targets, including the destruction of the World Trade Center towers on 11 September 2001. Documents and tape-recording from Al-Qaeda indicate that it is interested in acquiring weapons of mass destruction and using them on civilians. Although Al-Qaeda has not used these weapons, on 20 March 1995 the terrorist cult Aum Shinrikyo ('Supreme Truth') released sarin gas in a Tokyo subway, killing 12 people and injuring 5,500. The cult group also attempted, unsuccessfully, to spray anthrax spores over Tokyo (Cole 1996). In the autumn of 2001, someone – the culprit has not been caught – mailed anthrax spores to dozens of people in the Eastern US, killing four victims and sickening 20 others. The anthrax attacks caused a huge panic in the US as thousands of Americans took the antibiotic Cipro as a prophylactic measure. The idea that a terrorist group might one day use biological, chemical or nuclear weapons on civilian or military targets is not a paranoid fantasy; it is a real threat that should be taken seriously (MacKenzie 1998).

Given this social and political background, one can see why publishing information about how to make weapons of mass destruction could pose a significant threat to security. In the past two years there have been at least three papers published in prominent scientific journals that discussed methods and results pertaining to the genetic manipulation of deadly viruses. The papers were published while the US was assessing the threat posed by the use of smallpox as a bio-weapon and considering measures to address this threat, such as instituting a vaccination programme (Bozzette et al. 2003). In February 2001, the *Journal of Virology* published a paper that described the insertion of the gene for interleukin-4, an immune protein, into a mousepox virus. The researchers were trying to develop a method for rendering mice infertile. Instead, they developed a form of the virus that was much deadlier than the naturally occurring strain. The virus even killed mice that had been vaccinated against mousepox (Jackson et al. 2001). In June 2002, *Proceedings of the National Academy of Sciences* published a paper describing an experiment in which scientists formed a new smallpox protein complex, know as smallpox inhibitor of complement enzymes (SPICE), from a virus related to *Orthopoxvirus variola*, the virus that causes smallpox. Since the experiments also showed that the new protein deactivated human immune-system molecules C3b and C4b, a bio-weapon that delivered a genetically engineered smallpox virus with the new protein might be able to infect even people who have received the smallpox vaccine. The paper did mention, however, that it would be important to know how to disable the SPICE proteins (Rosengard et al. 2002). On 9 August 2002 (online version 11 July 2002), the journal *Science* published a paper on the creation of a polio virus by mail-ordering DNA from a private reagent company. The genetically engineered polio virus was capable of paralysing and killing mice (Cello, Paul and Wimmer 2002).

Many politicians and scientists objected to the publication of these papers and called for measures to censor biological research that poses security risks (Couzin 2002).

Several members of the US Congress introduced a resolution criticizing the publication of the polio-virus paper published in *Science*. In January 2003, the American Society for Microbiology (ASM), the National Academy of Sciences and the Center for Strategic and International Studies held a meeting in Washington, DC to discuss the censorship on biological research that poses security risks (Malakoff 2003). At the meeting, the editors of *Science, Nature* and a dozen other major journals said that they were already scrutinizing papers that raise security concerns, but that, so far, they had not rejected any. Several people at the meeting urged scientists to develop their own rules for self-censorship before governments start censoring scientific information (Atlas 2002).

This paper will not attempt to solve all of the complex problems concerning openness, freedom and national and international security. However, the paper will make a few general remarks about weapons of mass destruction that may help scientists, political leaders and policy-analysts focus on the important questions. First, there are at least four different ways of controlling the proliferation of weapons of mass destruction: (a) control of materials, (b) control of information, (c) control of scientists, and (d) control of governments or non-governmental groups. For some weapons, controlling the materials used to make the weapons will go a long way to preventing the proliferation. For example, it is difficult to obtain weapons-grade uranium or plutonium to make nuclear weapons. Countries that have signed the Nuclear Non-Proliferation Treaty co-operate with the International Atomic Energy Agency (IAEA) and other non-governmental organizations stopping the proliferation of nuclear weapons (International Atomic Energy Association IAEA 2003). The problem with this strategy is that the materials needed to develop biological or chemical weapons are not very difficult to obtain. Anthrax can be found in rotting carcasses, and most of the chemicals used to make some nerve agents can be purchased as pharmacies, grocery stores and agricultural supply stores.

Information with implications for national or international security is also very difficult to control. After World War II, the US attempted to keep nuclear secrets from the Soviet Union, but the Soviets soon learned how to build the bomb, as a result of their own research efforts and espionage. Today, university physics, chemistry and biology classes provide graduate students with enough information to build chemical, nuclear or biological weapons, and much of this general information is also available for free over the Internet. Indeed, the Internet has created tremendous challenges for controlling the flow of information by making it much easier to publish scientific information. Several decades ago, a handful of journals, government agencies and publishers would have been able to control most of the information related to constructing weapons of mass destruction. Today, almost anyone with a computer and a connection to the Internet can publish this information.

It may be somewhat easier to control scientists than it is to control materials or information. No country or terrorist group can develop weapons of mass destruction without the assistance of highly skilled scientists. After the fall of Saddam Hussein's regime, it will be very important to locate Iraqi scientists who may have been involved in Iraq's weapons programmes and employ them in peaceful activities. Although scientists are also difficult to control, it is certainly possible to influence scientists through employment opportunities, education and training in professionalism and ethics, and peer pressure.

Finally, it may also be possible to exert some control over the governments or non-governmental groups that want to develop weapons of mass destruction. Peaceful nations can exert economic, political and, if necessary, military pressure on countries that are seeking to develop weapons of mass destruction. Countries can also sign and monitor non-proliferation treaties. Although Iraq appears to have resisted a great deal of international pressure to relinquish its weapons programme, other countries, such as South Africa, have capitulated. Non-governmental groups are more difficult to control than governmental groups – how could one seriously negotiate with a terrorist organization such as Al Qaeda? – but it is also possible to exert some influence on these groups as well. For example, the Irish Republican Army, closely associated with the political organization Sinn Fein, agreed to cease its terrorist activities on 19 July

1997, after years of conflict and negotiation. One can also exert some influence or control over countries that sponsor or harbour terrorist groups, such as Afghanistan, which the US invaded, or Syria, which the US has recently criticized for its role in supporting terrorist groups such as Hezbollah.

The upshot of this discussion is that censoring scientific information is probably not a very effective way of preventing weapons or dangerous devices from entering the wrong hands. The most effective strategy is to control access to materials. When this strategy cannot work, it is far more effective to exert some influence over scientists, governments or non-governmental groups than it is to try to control the flow of information. On the other hand, one can acknowledge this conclusion and still maintain that stopping the flow of information to terrorist groups or rogue nations should be one part of a global strategy for curtailing the proliferation of weapons of mass destruction. Although other strategies may be more effective than censorship, this is still a useful strategy. Sometimes keeping some information secret for as long as one can is better than not keeping that information secret at all.

If countries decide to pursue censorship as means of stopping the spread of weapons of mass destruction, it is important to distinguish between scientific, self-censorship and governmental forms of censorship. There is an important legal difference between censorship by organizations that are non-governmental (or are not agents of the government) and censorship by the government. Many countries recognize a legal right to free speech. The US Constitution protects freedom of speech (Barron and Dienes 1999). In order for the US government to place restrictions on the content of non-commercial speech, the government must demonstrate that it has a compelling interest unrelated to the restriction of speech and that the method of restricting speech is the least restrictive method (United States v. O'Brien 1968). While national security is a compelling government interest unrelated to the restriction of speech, one might argue that censorship is not the least restrictive means of protecting this interest. For example, perhaps the government could allow publication in a forum with a limited audience. If one applies the strict scrutiny test to the controversial papers published in the last two years, it is not clear whether the US government would have had legal authority to stop publication of those papers. Another important concern in free-speech laws is vagueness and over-breadth: a law must not be so vague that people do not know whether it applies and not so broad that it deters legitimate speech (Barron and Dienes 1999).

If we consider censorship by non-governmental organizations, such as journals or professional associations, they would not have to face the legal challenges that governments would face, but they would still have to wrestle with moral questions. Journals and professional associations have an obligation to promote freedom, openness and other scientific values (Shamoo and Resnik 2002). On the other hand, they also have a moral responsibility to protect society from harm. In deciding how to respond to research with implications for national or international security, an organization must balance these competing values. Journals and professional associations may consider a number of different options, such as: (a) allow complete publication; (b) allow limited publication (e.g., restricted access to some parts of the publication); or (c) not allow publication. In choosing among these basic options, organizations should consider carefully the facts and circumstances of the case as well as the following factors relating to the nature of the security threat: (a) the gravity (or magnitude) of the threat, (b) the probability of the threat, (c) the imminence of the threat, (d) the preventability of the threat, and (e) the scientific and social value of the publication. If the threat posed by an article is grave, probable, imminent and preventable, and the publication has marginal scientific or social value, then a private organization would be justified in taking steps to stop its publication.

It is difficult to say whether any of the three controversial papers mentioned in this essay would meet all five of these conditions. The most dangerous paper was probably the one that described a mutated smallpox protein, because this paper showed people how to make a smallpox virus that might overcome human immunity (Rosengard et al. 2002). On the other hand, this paper could have some redeeming value in that it could help microbiologists and public-health experts learn how to develop immunizations against a smallpox virus that overcomes standard immunities. The irony of restricting dangerous publications is that the same information that could be used to make a deadly weapon could also be used for peaceful and productive goals.

Conclusion

As one can see, private interests and national-security concerns can pose significant problems for openness and freedom in scientific publication. Although this essay has discussed some potential solutions to these problems, more work needs to be done. Since these problems are global in scope, it is incumbent upon the scientists, policy analysts, concerned citizens and political leaders throughout the world to continue to find ways to safeguard openness and freedom in research while responding appropriately to emerging challenges to these values, and to co-operate internationally in the development of policies, practices and procedures.

References

Ackerman, W.M., 1998. Encryption: a 21st century national security dilemma. *International Review of Law, Computers and Technology*, 12 (2), 371-394.
Atlas, R.M., 2002. National security and the biological research community. *Science*, 298 (5594), 753-754.
Barinaga, M., 2003. Still debated, brain image archives are catching on. *Science*, 300 (5616), 43-45.
Barron, J.A. and Dienes, C.T., 1999. *Constitutional law*. 5th edn. West Publishing, St. Paul.
Blumenthal, D., Campbell, E.G., Anderson, M.S., et al., 1997. Withholding research results in academic life sciences: evidence from a national survey of faculty. *JAMA: the Journal of the American Medical Association*, 277 (15), 1224-1228.
Bozzette, S.A., Boer, R., Bhatnagar, V., et al., 2003. A model for a smallpox-vaccination policy. *The New England Journal of Medicine*, 348 (5), 416-425.
Campbell, E.G., Clarridge, B.R., Gokhale, M., et al., 2002. Data withholding in academic genetics: evidence from a national survey. *JAMA: the Journal of the American Medical Association*, 287 (4), 473-480.
Cello, J., Paul, A.V. and Wimmer, E., 2002. Chemical synthesis of poliovirus cDNA: generation of infectious virus in the absence of natural template. *Science*, 297 (5583), 1016-1018.
Cole, L.A., 1996. The specter of biological weapons. *Scientific American*, 275 (6), 61-65.
Couzin, J., 2002. Bioterrorism: a call for restraint on biological data. *Science*, 297 (5582), 749-751.

Dickson, D., 1984. *The new politics of science*. Pantheon Books, New York.

Freno, M., 2001. Database protection: resolving the United States database dilemma with an eye toward international protection. *Cornell International Law Journal*, 34 (1), 165-225.

Hilts, P., 1994. Philip Morris blocked '83 paper showing tobacco is addictive, panel finds. *The New York Times* (April 1 1994).

Hull, D.L., 1988. *Science as a process: an evolutionary account of the social and conceptual development of science*. University of Chicago Press, Chicago.

International Atomic Energy Association IAEA, 2003. *About the IAEA*. Available: [http://www.iaea.org/About/] (1 Apr 2004).

Jackson, R.J., Ramsay, A.J., Christensen, C.D., et al., 2001. Expression of mouse interleukin-4 by a recombinant ectromelia virus suppresses cytolytic lymphocyte responses and overcomes genetic resistance to mousepox. *Journal of Virology*, 75 (3), 1205-1210.

Kitcher, P., 2001. *Science, truth and democracy*. Oxford University Press, Oxford.

Kuhn, T.S., 1970. *The structure of scientific revolutions*. 2nd edn. University of Chicago Press, Chicago.

Longino, H.E., 1990. *Science as social knowledge: values and objectivity in scientific inquiry*. Princeton University Press, Princeton.

MacKenzie, D., 1998. Bioarmageddon. *New Scientist* (Sep 19 1998), 1-8.

Malakoff, D., 2003. Researchers urged to self-censor sensitive data. *Science*, 299 (5605), 321.

Marshall, E., 2001. Sharing the glory, not the credit. *Science*, 291 (5507), 1189-1193.

Marshall, E., 2003. The upside of good behavior: make your data freely available. *Science*, 299 (5609), 990.

Mauer, S.M. and Scotchmer, S., 1999. Database protection: is it broken and should we fix it? *Science*, 284 (5417), 1129-1130.

Merton, R.K. and Storer, N.W., 1973. *The sociology of science: theoretical and empirical investigations*. University of Chicago Press, Chicago.

Miller, A.R. and Davis, M.H., 2000. *Intellectual property: patents, trademarks, and copyright in a nutshell*. West Group, St. Paul.

Moreno, J.D., 1999. *Undue risk: secret experiments on humans*. W.H. Freeman, New York.

Nathan, D.G. and Weatherall, D.J., 2002. Academic freedom in clinical research. *The New England Journal of Medicine*, 347 (17), 1368-1371.

National Institutes of Health, 2003. *Final NIH Statement on sharing research data*. NIH, Bethesda. Notice no. NOT-OD-03-032.
[http://grants1.nih.gov/grants/guide/notice-files/NOT-OD-03-032.html]

National Research Council NRC, 2003. *Sharing publication-related data and materials: responsibilities of authorship in the life sciences*. The National Academies Press, Washington DC.
[http://books.nap.edu/books/0309088593/html/index.html]

Nebbia v. New York, 1934. 291, U.S., 502 (1934).

Olivieri, N.F., 2003. Patients' health of company profits? The commercialisation of academic research. *Science and Engineering Ethics*, 9 (1), 29-41.

Resnik, D.B., 1998. *The ethics of science: an introduction*. Routledge, London. Philosophical Issues in Science.

Rosengard, A.M., Yu Liu, Zhiping Nie, et al., 2002. Variola virus immune evasion design: expression of a highly efficient inhibitor of human complement.

Proceedings of the National Academy of Sciences of the United States of America, 99, 8808-8813.
Shamoo, A.E. and Resnik, D.B., 2002. *Responsible conduct of research.* Oxford University Press, Oxford.
United States v. O'Brien, 1968. 391 U.S. 367 (1968).
Wadman, M., 1996. Drug company 'suppressed' publication of research. *Nature,* 381 (6577), 4.
Ziman, J., 2002. *Real science: what it is and what it means.* Cambridge University Press, Cambridge.

6b

Comments on Resnik: Some recent challenges to openness and freedom in scientific publication

Tiny van Boekel[#]

The paper points out the dangers of industry-sponsored research, but I would also like to draw attention to the opportunities, namely involving the industry in the research in order to create an interaction between science and society. This will primarily concern pre-competitive research; at least in the Netherlands, this kind of co-operation is actually stimulated. There are possibilities to limit the dangers, e.g. by delaying (but not prohibiting) publication for a period of, say, 6 months. As for intellectual property rights, written arrangements can be made before the research starts. It is also in the interest of the industry to educate science students in such a way that they are useful to the industry once they have graduated.

It seems to me that the distinction between public laboratories searching for truth and private labs helping the company make a profit is an artificial one. First of all, the search for truth is debatable, as public labs have their priorities as well. For instance, in the field of food science and nutrition, reality is so complicated that results will never be 'true'. Moreover, there is a drive to publish, not because of wanting to reveal the truth but because of the scientist's career, and the publication may include results and data that are not complete or are based upon research that was not properly conducted. Furthermore, editors of scientific journals are not keen to accept 'negative' results, e.g., results on drugs or foods that are shown not to have the anticipated effect. So, there is a bias in what is published, also from public labs, and the idea of truth becomes blurred. Finally, there is nothing wrong in doing science to help a company make a profit, as long as the science is carried out properly and no evident harm is done to society. Some companies have the policy to publish (some of) their obtained scientific results so that they can be seen as full and competent players in the scientific arena.

I would like to make a plea for the publication of raw data (on the internet for instance), whether or not they come from public or private labs. These data should be in the public domain and they should be available free of charge. Only for processed data a certain amount of money may be asked.

As for the notion of 'dangerous' research (i.e. research that results in data that could be used for instance for military purposes or by terrorists), I would like to remark that there is no dangerous research in my opinion. Results or data only become dangerous in a certain context, and I would strongly oppose self-censorship of scientists who withhold information or authorities that control scientific information just because it might be used in a wrong way. I do not want to deny the possible threats, but the information may also be used in a very positive way. In my own area of food science, there are quite a few positive developments made possible by research that was originally intended to serve military interests.

[#] Product Design and Quality Management Group, Department of Agrotechnology and Food Sciences, Wageningen University, The Netherlands. E-mail: TinyvanBoekel@wur.nl

ETHICS OF ANIMAL RESEARCH

7a

Research ethics for animal biotechnology

Paul B. Thompson[#]

Animal biotechnology can be broadly categorized as encompassing the asexual reproduction of animals through cloning, and genetic transformation of animals through the manipulations made possible through recombinant DNA. The character and methods of such manipulations include the creation of 'knockout' animals intended to study gene function on the one hand, and also the insertion of genes originally identified in other species, or, colloquially, genetic engineering, on the other. This definition will clearly change and grow with theoretical and technological developments in genomics and systematic biology, but for the time being cloning and genetic transformation represent the main foci of animal biotechnology for the purpose of research ethics.

Bernard Rollin's 1986 paper *"The Frankenstein Thing"* articulated two ethical principles for animal biotechnology. One was the principle of conservation of welfare, to wit, that applications of biotechnology should result in animals that are no worse off with respect to suffering and frustration than traditionally bred animals. The other was his view that biotechnology should be used to make animals less likely to suffer in the various settings that they are to be used. This could be called the 'improvement of welfare' principle, and Rollin understood it to entail one uncontroversial and one very controversial conclusion. The uncontroversial conclusion was that there is nothing wrong with using biotechnology to address problems of animal health. The controversial one is that there is nothing wrong with using biotechnology to produce animals less capable of experiencing the suffering that humans inflict upon them.

Rollin's original article addressed the central question of research ethics for use of animals: what are researchers' responsibilities with respect to animals that they use in research? Many of the subsequent reactions to animal biotechnology have been less clearly relevant to this question, and many have ascended to a much broader level of generality, questioning, for example, the moral status of transgenic animals in general (Balzer, Rippe and Schaber 2000) or society's general responsibilities to transgenic animals (Bovenkerk, Brom and Van den Bergh 2002). In this paper I want to return to the narrower questions of research ethics. The approach that I will outline is a form of pragmatic bioethics (Keulartz et al. 2002). I begin with an overview of debate over animal biotechnology, followed by a brief discussion of animal welfare and research ethics. I argue that the key research-ethics questions demand a scientifically informed approach to animal welfare, which in turn demands an understanding of the interpenetration between ethics and animal-welfare science. I conclude with a discussion of how research-ethics committees can approach the evaluation of animal biotechnology in a more ethically satisfactory manner. My treatment of these issues reflects my background and main research interests with traditional livestock species

---

[#] Department of Philosophy, Michigan State University, East Lansing, MI, USA. E-mail: thomp649@msu.edu

that are intended for the production of animals for meat, milk and animal by-products, rather than mouse biotechnology where the focus has been on developing models for human disease.

Welfare and biotechnology: background

There is ample documentation of the public's interest in biotechnology's impact on animal welfare. In one of the early studies, animal biotechnology was found to be thought ethically problematic by a greater percentage of respondents than genetic engineering applied to human beings (Hoban and Kendall 1993). One British study found that public was more accepting of animal genetic engineering for biomedical research than for food production (Sparks, Shepherd and Frewer 1995), and another study by the same group found that the British public interpreted impact on animals as an ethical issue, rather than as an effect bearing on risk (Frewer, Howard and Shepherd 1997). There is evidence that respondents find animal cloning or transformation to be particularly problematic, as distinct from those who have general objections to the use of gene technologies as well as those who associate animal biotechnology with detrimental impact on animal welfare (Durant, Bauer and Gaskell 1998).

As early as 1992, the National Agricultural Biotechnology Council (NABC) held a meeting on animal biotechnology where animal welfare was a main focus of discussion. The consensus workshop of that meeting called for empirical research on the welfare of transgenic animals (McDonald 1991). NABC endorsed reasearch on both scientfic and philosophical dimensions of the animal-welfare question for a second time in 1995 (Thompson 1998). Similar recommendations were made in the Polkinghorne report (Ministry of Agriculture 1993) and by two subsequent committees in the United Kingdom (Bruce and Bruce 2000). A workshop of the European Centre for the Validation of Alternative Methods (ECVAM) published detailed recommendations for conducting research on the welfare of transgenic animals, though many were directed particularly toward biomedical applications such as the development of animal models for the study of human disease (Mepham et al. 1999).

There have also been many conceptual and philosophical papers on animal welfare and biotechnology. Rollin's 1986 paper presented an analysis that argued against the belief that simply introducing novelty into the genome could be construed as a form of harm. Instead, Rollin suggested that the entire ethical significance of animal biotechnology resides in risk to humans, environment and to the animals themselves. Rollin has extended but not substantially changed this analysis in subsequent writings, including his 1995 book. Less favourable philosophical viewpoints on animal biotechnology have been offered by Fox (1990), Linzy (1990), Verhoog (1992), Rifkin (1995), Ryder (1995) and Holland (1995). A complete and in-depth debate of the conceptual issues can be found in the contributions by scientists and philosophers to Holland and Johnson's Animal Biotechnology and Ethics (1998). Thompson (1997; 1999) and Appleby (1998) have published critical discussions of this literature that essentially support Rollin's analysis. De Cock Buning (2000) has argued that these less favourable approaches to understanding the moral issues behind genetic engineering of animals imply a deontological evaluation of the process of gene transfer and exclude the relevance of consequentialist norms, such as Rollin's principles, that emphasize the outcome with respect to animal welfare. He concludes by expressing the hope that some middle ground could be struck between these approaches, but does not offer a principled way forward.

Expressions of public concern and ethical relevance notwithstanding, there are relatively few published articles discussing empirical studies of the welfare of genetically engineered or cloned farm animals. Van Reenen and Blokhuis (1993; 1997) report that adverse impact on the welfare of the transgenic cattle that they studied was limited to the experimental stage in which pre-implantation embryos were manipulated in vitro. Hughes et al. (1996) report no significant differences between transgenic and control sheep. Jaenisch and Wilmut (2001) report abnormal development in cloned sheep, but Lanza et al. (2001) report no abnormalities in a group of cloned cattle. A survey article by Mepham and Crilly (1999) extracts data from several research reports on transgenic farm animals that were not designed as studies of welfare to support the conclusion that transgenic animals may suffer from adverse welfare when compared to non-transgenic livestock. A recent survey article by Heap and Spencer (2000) cites anecdotal reports of adverse impact associated with both transgenics and cloning, but does not cite any empirical data.

In sum, public attitudes toward the ethical acceptability of animal biotechnology depend heavily on its impact on animal welfare. The need for empirical research on such impact has been endorsed by a number of groups representing scientific interests, as well (National Research Council NRC 2002). Given the relative paucity of published research on the topic, it is difficult to make any confident assertions about the relationship between welfare and biotechnology. This places researchers who must evaluate the ethics of work using biotechnology in a difficult position, the nature of which becomes clear when we frame the questions that must be asked within current approaches to research ethics.

Animal use in a research-ethics context

Research ethics takes an immediate concern with the conduct of research, though broader questions about the impact of science on society may also be appropriate. A procedural approach to the ethics of using animals in research has evolved within the science community over the last several decades. A procedural ethic is one in which the ethical justifiability of a particular course of action is tied to that course of action having been endorsed by a well characterized decision procedure. In contrast to procedural approaches, substantive ethical theories stipulate general principles or norms (such as the categorical imperative or the utilitarian maxim) and then interpret justification in terms of consistency with the stipulated principles or norms (see Russow 1999). The procedural nature of animal ethics within the context of scientific research is evident in the reliance on committee approval processes that are now commonplace across the globe.

Although the details for animal-ethics committees vary from country to country, the general approach to animal ethics within research institutions has been to require that researchers develop a protocol describing how animals will be used in a proposed experiment. These protocols are then reviewed by a committee which must decide whether or not the use of animals is acceptable. In most cases, this committee or a complementary one also has responsibility to ensure that actual practice is consistent with what has been approved in the protocol. The committee(s) itself (themselves) may be subjected to additional forms of oversight from authorizing or regulatory bodies. Thus, in the United States, institutions conducting animal research are required to constitute an internal Animal Care and Use Committee (IACUC) that will review and approve protocols, as well as inspecting and certifying that animals used for research in the institution are in fact being used as indicated. Institutions and their IACUCs are inspected by the USDepartment of Agriculture, which has statutory authority to ensure that the provisions of the Animal Welfare Act are being upheld (National Research Council NRC 1996; Office of Laboratory Animal Welfare OLAW 2002).

The procedural IACUC-style approach to animal ethics ensures that animal interests are, at some level, being considered in the planning and conduct of research, at the same time that it allows for considerable flexibility in terms of specific administrative approaches. Not only will the administrative style of an institution result in different approaches, but also the nature of research being done at the institution may mean that different types of questions are appropriate. A drug company may be using rather different animals and conducting rather different kinds of studies from an agricultural college or a wildlife research institute, for example. It must be admitted that nothing in the procedural approach itself guarantees that the right ethical questions regarding animal use are being considered or that they are taken with an appropriate level of seriousness. Thus, the nature of the approval and oversight procedures at any given institution will reflect the culture and values of that institution (Rowan 1990). One interesting and ironic aspect of the culture within USinstitutions is a reluctance to using the word 'ethics' to describe what is essentially an ethical process, thus Americans have 'care and use' committees, while the rest of the world is more comfortable with animal-ethics committees (Jennings and Miller 2000; Marie et al. 2003).

Nevertheless, the evolution and development of the IACUC/ethics-committee approach has produced a fair amount of global agreement on the principles that tend to get used within the committee structure. First and foremost, animals count and scientists have ethical responsibilities for their care. Second and of almost equal importance, at least some human use of animals for research purposes is acceptable. Research ethics for animals remains fundamentally at odds with philosophical views that do not accept this fundamental tenet. Research ethics is thus committed to a broad interpretation of the view that the benefits or goals sought through research can, in some cases at least, outweigh or override harms to the animal research subjects on which that research is conducted. Third, research is acceptable because the committee members have approved it, either through a process of formal voting or a consensus procedure, and not because it conforms to some philosophical standard. (Donnelley 1990; Jennings and Miller 2000; Heitman 2002).

The third tenet is actually quite important, for it stipulates the procedural nature of the norms guiding animal research and creates a pluralistic structure for the application of substantive principles regarding animal use. A number of different standards, patterns of evaluation or principles of justification are available, and each member of the committee may apply their own judgment and standards to a protocol or policy. They are voting to approve or deny a protocol, not to decide a comprehensive philosophical view of animal ethics. Furthermore, the procedural pluralism of animal-research ethics qualifies the first two tenets in the following sense: although the benefits of research can in principle outweigh or offset harms to animals, this offsetting should not be thought of as involving a classical cost–benefit style of justification through optimization. Some members of a committee may be thinking that way, but others may not.

Beyond this, the details of animal research ethics can become almost overwhelming. One key question concerns the make-up of committees. Current U.S.

procedures require both a non-scientist and a member not affiliated with the institution conducting the research. These requirements can be interpreted to suggest that public concerns such as those noted above do indeed have a place within the procedural ethics of animal research. Furthermore, committees clearly need veterinary and animal-welfare expertise, as well as sophistication in experimental design. The ethical significance of these areas of expertise is sometimes summarized in terms of the 'Three R's'. Scientists should seek to reduce the number of animals being used. In practice this means that protocols are reviewed carefully with respect to whether experiments are decisive, including whether the number of animals is sufficient to generate data that will meet statistical requirements, but not exceeding the number in ways that inflict needless suffering or inconvenience on animal subjects. This requires expertise in statistics and research design. Scientists should seek to replace animals when possible, using computer or tissue models, perhaps, but also using species that have less complex and demanding cognitive needs. Replacement requires expertise in basic theory and in alternatives to animal use. Finally scientists should develop approaches that refine whatever adverse impacts may be unavoidable. Pain and discomfort should be treated with appropriate analgesics, husbandry and, in the extreme case, euthanasia (Russell and Burch 1959; Orlans 1990). Interpreting refinement within the context of an animal-ethics committee review involves standard veterinary expertise, but increasingly it also involves application of applied animal-welfare science.

Institutional animal-ethics committees should be constituted with these three broad areas of expertise in mind. Such committees typically involve representation from disciplines and research programmes conducting animal research, and as such are likely to include members with the requisite knowledge for reduction and replacement. Membership from attending veterinary staff is also required in the U.S., but veterinarians may or may not be cognizant of welfare science. None of the needed areas of expertise explicitly involve philosophy or ethics; hence one might wonder what role ethics would have in conducting the business of animal-ethics committees. But ethics and welfare science are intertwined. Thus it is useful to examine this element of the requisite ethical expertise for animal-research ethics more carefully.

Animal-welfare science

The science of animal welfare has grown considerably over the last two decades. It can be described generally as a blending of longstanding approaches to the assessment of animal health along with studies of animal behaviour and most recently cognitive and neuroscience methods for making comparative assessments of how animals fare with respect to different husbandry practices and environmental conditions. Animal-welfare science thus builds upon standard and relatively non-controversial veterinary approaches to animal health. Behavioural and cognitive research on animals grew out of the attempt to understand and model the functioning of individuals and animal populations in the wild. As it became evident that mechanistic, instinctual and physiological accounts of animals were unable to account for the complexity of animal behaviour in the wild, the application of cognitive ethological methods to animal welfare was a natural move (Bekoff 1994). However, much of the progress that has been made in animal-welfare science became possible only when researchers adopted rather pragmatic attitudes toward a number of conceptual and methodological problems that constrain research on animal welfare. Any application of science to the study of welfare issues associated with biotechnology will be subject to these constraints. In addition, cloning and genetic transformation introduce additional complications into the basic problems of conceptualizing and measuring the welfare of farm animals.

Animal welfare is a multi-attribute, multi-disciplinary phenomenon

Considerable energy was expended during the early years of research on animal welfare debating the basis for an attribution or evaluation of comparative well-being. In fact, as with human well-being, there are a number of attributes to animal welfare, and attempts to formulate a reductive approach have not proved successful. Physiological indicators of health, such as growth rates and the absence of recognized disease etiologies are clearly relevant to welfare. Ongoing or cumulative experiences of pain and stress are elements of welfare, though it is doubtful that total absence of pain or stress would be consistent with acceptable levels of welfare in humans or livestock. Behavioural studies and evolutionary genetics suggest that animals of a given species have instinctual or functional needs for varying degrees of social contact, spatial orientation and sexual activity, as well as for performing certain behaviours. The scientific methods available to assess each of these attributes tend to be associated with different sub-disciplines in the animal sciences. Each of these sub-disciplines employs different experimental procedures that traditionally characterize dimensions of welfare not amenable to measurement by those procedures as elements that should be controlled through good experimental design. Thus one researcher's dependent variable is another's confounding factor. Animal-welfare researchers have thus faced the difficult conceptual problem of integrating methods and theories developed in quite disparate areas of science (Appleby 1999).

Welfare is an inherently normative concept

The very idea of welfare implies a normative valuation associated with a given state of affairs. To associate a given state of physiological functioning with an animal's welfare is to judge, at least implicitly, that this state is indicative of the animal's doing well or poorly. Nature, it could be said, is indifferent with respect to these various states; so to characterize them as elements of welfare is to interpret these states within a framework of valuation. Some of these evaluative elements are quite uncontroversial. For example, a disease process resulting in the death of an organism would readily be characterized as inimical to that organism's welfare. Yet such a characterization implies that life is better than death, and this, in turn, implies a perspective (perhaps that of the organism itself) from which such a value judgment can be made. Being a complex blending of multiple attributes, assigning value to a particular physiological or behavioural state is in fact often quite controversial. Such valuations are particularly susceptible to well-known fallacies. Anthropomorphism consists in presuming that a state or behaviour valued by human beings would similarly be valued by animals of another species. The naturalistic fallacy consists in presuming that a state or behaviour characteristic of animals in a 'natural' environment is, for that reason, particularly good, valuable or indicative of welfare. This fallacy is compounded by difficulties in ascertaining what environments would count as natural for domesticated animals. The ecological fallacy consists in presuming that what is good for the individual organism is good for the species, or vice versa.

There is a further fallacy that might be called the positivist fallacy. This consists in presuming that because the normative dimension of a phenomenon is elusive, ambiguous and subject to erroneous classification, it not only can but perhaps should be ignored altogether. In its most extreme version, the positivist fallacy consists in the performative contradiction of asserting that science should be value-free. Of course, a value-free science could not include the normative assertion that it should be value-free, hence the contradictory nature of making such an assertion. According to numerous observers of the animal sciences, the positivist fallacy has played an influential role leading to the relative neglect of animal welfare in traditional agricultural research disciplines (Kunkel 2000; Rollin 1989). In fact, as a domain of applied science dedicated to increasing yield, productivity and the enhancement of sensory, nutritional and economic value of food commodities, the agricultural sciences presume elusive and occasionally contested value judgments in their most basic goals and concepts. The neglect of animal welfare thus involves a rather selective and arbitrary application of the positivist fallacy within the context of agricultural research.

Some normative approaches to animals are more conducive to scientific studies of welfare than others

David Fraser (1999) describes two broad approaches to animal ethics. Type I is characterized by Peter Singer and Tom Regan, two philosophers prominently associated with animal advocacy. This view is characterized by an attempt to develop a single ethical principle that would be used to derive a normative valuation for individual animal lives. Type-I animal ethics does not discriminate according to taxonomic differences among animals, and presumes that a single normative criterion – generally associated with sentience – can be used to derive standards for human conduct oriented to animal welfare. Type-II animal ethics, which Fraser associates with a number of less well-known philosophers, takes a more inductive philosophical approach that attempts to specify principles for ethical treatment of animals that is sensitive to different species cognitive and behavioural capabilities, as well as to different practices of husbandry that might emerge given particular economic, technological and political preconditions. Fraser argues that Type-II animal ethics is more conducive to application within animal welfare science, and that given the inherently normative basis of welfare, is in fact essential to the development of coherent research programmes in animal-welfare science.

Type-II ethics can be characterized as a form of philosophical pragmatism (Thompson, forthcoming). Here, the philosophical starting point is to review why the situation at hand might be thought problematic. As already indicated, animal-welfare science is most relevant to problems of refinement: finding ways to mitigate the pain and discomfort suffered by experimental animals. Here, standard veterinary indicators of animal health can be used along with behavioural cues, such as stereotypies and avoidance behaviour, to make assessments of animal preferences. These welfare indicators are combined with an attempt to assess cognitive capabilities and functional or instinctual drives typical of a species. In addition, both the evolutionary history of the species (including domestication events, for domesticated species) and the ecological history of environmental conditions in which founder animals for current populations have lived are used to provide a background framework for comparing a protocol as designed to possible alternatives. Although this inductive and comparative approach does not yield criteria that permit one to judge given measures of welfare to be unambiguously acceptable or unacceptable, a pragmatic approach to ethics does facilitate amelioration of animal-welfare problems within the parameters set by experimental design.

Welfare science and animal biotechnology

Given a pragmatic and procedural approach to animal-research ethics, it is now time to explore briefly a few of the questions and approaches that are most relevant to animal biotechnology. The Principle of Conservation of Welfare that Rollin (1986; 1995) proposed as the standard of welfare for agricultural biotechnology states that, all things considered, transgenic and cloned animals should not be worse off than founder animals or other animals of the same species used for similar research purposes. Rollins' Principle of Conservation of Welfare permits applications of biotechnology that make animals better off than the comparison group. It should be possible to apply animal-welfare science to transgenic and cloned animals in collecting a variety of standard veterinary measures of health and development, augmented by behavioural studies. Such data could be compared against standard ranges of value considered to be typical of healthy animals, and also to data taken for non-transgenic, non-cloned animals in control groups. This comparison would provide the basis for an empirically informed assessment of how these animals fare in comparison to conventionally bred animals kept under comparable circumstances. At the risk of stating the obvious, I will explore the ramifications of such an approach in more detail.

If the procedures used in gene transfer or cloning are themselves the source of detrimental (or positive) effects on animal welfare, we would expect that welfare indicators for a sufficiently large and random sample of cloned and transgenic animals would diverge from those of conventional livestock to a statistically significant degree. Given the standard conditions of variability associated with measurements of welfare, we should not expect that all data will provide an unequivocal indication of any one result, at least until a number of studies have been conducted. The current situation, in which relatively few studies have been conducted, does not provide a sufficient basis for making a scientifically informed judgment. The existence of at least a few studies indicating no significant impact on scientifically measured indicators of welfare does show that techniques of transgenesis and cloning do not necessarily lead to detrimental impacts on animal welfare, and we should not underestimate the significance of even this limited result. It does provide a temporary basis for animal-ethics committees to use when evaluating studies involving cloned and transgenic animals. Nevertheless, any scientifically informed judgment about the effect of biotechnology on animal welfare will require a significantly larger number of studies.

There are confounding factors that clearly apply to transgenic animals. The choice of genes used in transformation may substantially affect the measurements that are associated with specific welfare indicators. It is generally accepted that a subclass of possible genetic modifications would be expected to produce detrimental results for animal welfare. Animals developed as models of human disease, for example, would not be characterized as successful transformations unless they exhibit a given disease in a large percentage of individuals, hence they would be expected to perform less well than a typical individual with respect to welfare indicators. As a second example, it is now widely hypothesized that the transformation associated with the famous Beltsville pigs – increased production of growth hormone – was contrary to animal health. In such cases, successful incorporation and expression of the transgene can be expected to have a detrimental impact on animal welfare. Thus, there will be a subpopulation of transgenic animals for which the welfare comparison is unfavourable, but which do not provide evidence that the techniques for producing transgenic animals are responsible for the negative impact on welfare, as distinct from the expected functioning of the transgene. It may not be possible to discern which transgenes have detrimental effects without experimentation. There may, of course, prove to be similar confounding factors for cloning, particularly if cloning is practiced primarily on a subpopulation of founder animals that is atypical with respect to other welfare indicators.

A second set of possible confounding factors are derived from welfare indicators associated with reproductive health. Both transformation and cloning involve in-vitro manipulation of embryos and blastocysts. In-vitro fertilization and implantation is itself associated with reduced levels of reproductive success, and the manipulations associated with biotechnology can be expected to lower the survival rate of blastocysts, as well as live births. If such data are incorporated into comparisons between conventional in-vitro fertilization and either transgenic or cloned animals, it will almost certainly skew the resulting assessment against animal biotechnology. Comparison with in-vivo fertilization will be even worse. Yet however issues associated with reproductive success are finally evaluated, keeping these issues separate from those associated with live-born transgenics and clones would appear to be warranted. For the near term, all transgenic and cloned animals must be regarded as experimental. It is reasonable to expect that current rates of reproductive success may not be typical of those that would be associated with more mature technology. As such, meaningful welfare comparisons of reproductive-success data may be premature. Furthermore, it is far from clear how to apply scientifically measurable indicators of welfare to prenatal and even early neonatal individuals. If such comparisons are to be made on a scientific basis, there will need to be a corresponding development in the conceptualization and testing of measures for animal welfare during the earliest stages of life.

There are a number of practical issues that will arise in conjunction with any attempt to apply animal-welfare science to biotechnology. Once concerns the specific indicators and measures that would be deployed in any attempt to make comparative assessments of welfare. The ECVAM working group attempted to specify a number of fairly specific indicators in their 1999 report focused on laboratory animals (Mepham et al. 1999). While it may be useful for the NRC committee to provide some general guidance on welfare measures, it is questionable as to whether any specific set of indicators would be appropriate for all transgenic and cloned animals. One can expect that there will be significant differences depending on species and the nature of a genetic transformation, not to mention differences that depend on the number of animals and the experimental conditions in which measurements can be taken. The ECVAM group also recommended that all experiments involving transgenic animals involve an animal-welfare component. Yet data beyond the most rudimentary indicators of health and development might not be particularly indicative of welfare for transgenic and cloned animals in certain cases. Collection of data and operation of controls may, in some cases, involve costs that would not be justified by the scientific results that could be expected to be obtained. However, these practical difficulties should not be allowed to result in a continuation of the current paucity of meaningful scientific data on the welfare of transgenic and cloned animals.

Conceptual and philosophical issues

Although it should possible for animal-welfare science to make a significant contribution to our understanding of welfare issues associated with animal biotechnology, there are a few conceptual and philosophical issues that do not appear to be resolvable on scientific grounds. Many issues revert back to the difference between Fraser's Type-I and Type-II (or pragmatic) animal ethics. It is not clear that every approach Fraser has classified under the Type-I heading would be opposed to animal biotechnology. Singer's utilitarianism, in particular, might be applied in a manner that would find some applications of biotechnology to be morally permissible. Nevertheless, there is a growing philosophical literature that opposes genetic engineering and cloning on grounds that are not particularly amenable to the empirical, inductive and comparative approach of animal-welfare science. There are two broad types of argumentation that are relevant. One associated with Fox (1990), Verhoog (1992) and Rifkin (1995) puts forward the idea that animal genomes possess a form of integrity such that modification is in every case a violation of integrity and morally forbidden. The second involves the claim that modification is not consistent with the animal's own good, and as such involves a form of disrespect. Ryder (1995) and Holland (1995) offered versions of this argument, and a recent article by Balzer, Rippe and Schaber (2000) uses a similar claim to suggest that biotechnology is in all cases inconsistent with the dignity that should be accorded to all animals. Balzer et al. note that current Swiss law requires respect for the dignity of creatures and argue that this forms a basis for opposition to biotechnology on legal grounds.

Arguments that stress the genetic integrity of animals may be subject to a scientific evaluation. At a minimum, it would be useful to debate whether the notion of genetic or genomic integrity can be supplied with any scientifically grounded basis. However, it seems most likely that this will be a topic for geneticists, evolutionary theorists and possibly ecologists. Animal-welfare science will play at most a supporting role to these other disciplines in biology. It may appear that welfare science could be brought to bear on arguments emphasizing an animal's own good. However, these arguments emphasizing dignity suggest that the underlying ethical problems follow from a viewpoint that interprets animals as of purely instrumental value for human uses. As such, this philosophical approach follows not from an empirical assessment of how animals fare in particularly situations, but from an assessment of the human attitudes that lie behind animal use. Though the philosophical merits of both arguments should be analysed and debated within the context of animal ethics, neither will be deeply affected by empirical considerations derived from animal-welfare science.

A more interesting and relevant set of conceptual issues arise in connection with genetic modifications that improve the well-being of animals both in biomedical research applications and in livestock-production settings by changing their capacities to experience various forms of stress or deprivation. This issue has been debated extensively in connection with Bernard Rollin's discussion of whether it would be unethical to change an animal's telos. 'Telos' is a term Rollin adapted from Aristotle's philosophy to summarize the behavioural, functional and cognitive drives that are most deeply associated with the welfare or well-being of animals of a given species. Thus, nesting behaviour or various forms of social grouping would be characteristic of the telos for some animals, but not for others. Rollin (1995) has argued that not only is there no reason why it should not be permissible to modify an animal so that certain drives or needs characteristic of founding individuals are absent in the modified progeny, but also that for animals that intended to mimic painful disease conditions, such forms of modification may be ethically mandatory as a form of refinement. In some cases, arguments stressing integrity or dignity have been an attempt to find some grounds on which to oppose this result.

The modification of functional needs and drives also leaves open the possibility that one might relieve welfare problems associated with livestock production by using genetic engineering to produce a new breed that would not be vulnerable to stressors that lead to these problems. This is not a possibility that is unique to biotechnology. Conventionally bred species of blind hens display lower levels of stress in typical confined layer production systems. Standard measures of welfare would thus indicate the use of blind hens would be preferable to sighted hens. Yet few are willing to follow this application of welfare logic (Sandøe et al. 1999). This suggests that at least some applications of biotechnology that would be supported by the comparative analysis of animal-welfare science will be found questionable on ethical grounds. This is a curious result. It seems unlikely that people would reject a modification that made livestock less susceptible to a production-related disease, such as mastitis, yet the judgment that it is not acceptable to solve welfare problems by making animals less susceptible to stress is widely supported by anecdotal analysis. This may be more of a philosophical than a scientific problem, yet it does suggest that there may be some possible areas of ethical, conceptual and philosophical ambiguity where the application of animal-welfare science will be less than straightforward.

Meeting public expectations

These conceptual issues bring us back to the broader context of public interest in the ethics of animal biotechnology. The existence of ethical puzzles about using biotechnology to pursue the goals of refinement suggests that there is an additional way in which animal biotechnology is problematic. Some strategies for mitigating animal suffering may be at odds with our culturally based views about what it is acceptable to do. Animal-ethics committees intently focused on achieving the goals of refinement may find themselves at odds with the broader public if they sanction transformations that would substantially alter the basic characteristics of a species. Rollin's suggestion that it would be a good thing to develop a strain of decerebrate lab animals incapable of feeling pain is unlikely to win wide public approval, at least over the short run, and even less radical uses of biotechnology in the service of refinement can certainly run afoul of ethical norms that have been articulated in terms of respecting the intrinsic value of research animals. It is in responding to this tangle of issues that the pragmatism of my approach becomes most evident.

First, the *sine qua non* of pragmatism is its unrelenting focus on the problem at hand. Other philosophical schools have tended to presume that deep and enduring philosophical disputes such as the irreconcilability of deontology and consequentialism or the puzzle of *Moralität* and *Sittlichkeit* lie at the heart of every moral issue. Pragmatists try to avoid prejudging the philosophical issues, and are especially attentive to what Todd Lekan (2003) calls 'determination problems', where the exact nature of what is at issue is precisely what needs philosophical attention. In the present instance there are at least two problems that are deeply entangled. One is that we are not at all sure of what the right thing to do is in the cases that have been reviewed in the preceding section. This problem is grounded in the lack of clarity regarding ethical responsibilities that scientists owe to research animals. The examples of blind hens and decerebrate mice indicate a puzzle over the way that researchers' duties to animals ought to be discharged, and there is every reason to believe that no convincing solution to it is on the near horizon. As Bernard Gert argues in his contribution this volume, the issues are beyond the reach of common morality. The other problem is that animal-ethics committees need some guidance on how to proceed on a provisional basis when there is a high degree of confusion and moral uncertainty. This problem is particularly acute in light of the significant degree of public interest in the ethics of agricultural biotechnology, and it is useful to understand this problem as grounded not in responsibilities to animals, but in the responsibility of scientists to the broader public.

Scientists are accorded special roles in society in order to carry out their work, and in exchange the public is entitled to expect that scientists are conducting their business in a thoughtful and deliberative manner. Animal-ethics committees are an institutional response to this public expectation. They address both substantive concern with the care and treatment of research animals and also the need for public accountability with respect to the care and treatment of animals. The procedures and records of committee activities provide a basis for the public to see whether researchers are taking their responsibilities to animals seriously. Looking toward institutions that can cope with vagueness, indeterminacy and a plurality of relevant norms and perspectives is also a characteristically pragmatic response. Pragmatists tend to favour procedural norms in such instances, especially when procedural approaches have the capacity to provide some basis for taking action in the short term without foreclosing the possibility for revisiting an issue (or taking a different stance with respect to apparently similar cases) as experience and learning accumulate. Pragmatists also see normative standards as products of social interaction; hence the institutional and procedural orientation of animal-ethics committees represents a characteristically pragmatic response to the problem of public expectations.

Members of the public can appreciate the conceptual and philosophical difficulty of issues such as those that have been discussed above. Public opinion about the way that such questions should be settled will almost certainly be as diverse and divided as that of the philosophers themselves, and research-ethics committees cannot be expected to settle the issues, especially not in the context of reviewing protocols. However, it is reasonable to expect that even these difficult issues are being taken seriously. It would be ethically problematic if researchers used the uncertainty and philosophical difficulty of determining what our responsibilities are with respect to cloned or transgenic animals as an excuse for ignoring the potential problems altogether. Hence there is an ethical responsibility both to grapple with these issues and to be able to demonstrate to the public that researchers have not taken a cavalier attitude with respect to them. This is not to say that the philosophical puzzles of duties to research animals are resolved. What can be said is that the puzzles are being taken seriously, and that, by implication, responsibilities to animals are being taken seriously.

A pragmatic approach to procedures for addressing researchers' responsibilities to research animals should produce an amelioration of problems associated with animal use, and should be oriented toward public, institutional learning that extends beyond the particular individuals who happen to sit on committees at any given time. With this in mind, it is possible to offer the following suggestions as part of the approach for IACUCs and animal-ethics committees to take in response to the challenges posed by animal biotechnology.

At a minimum, committees should have records that indicate what research is being done with cloned and transgenic animals. These records should describe cloning methods and/or the nature of the genetic transformation in functional terms. This is a minimal requirement for being able to monitor and document whether cloned and transgenic animals suffer unusual health or well-being problems. If record keeping is tightly linked to protocols, this means that generic protocols for performing a broad class of biotechnology experiments will not be acceptable.

Since there are few public data on the general effects of biotechnology on animal health and well-being, committees should encourage the collection, sharing and publication of such data whenever doing so is compatible with larger research purposes and institutional policies.

Particular experiments involving genetic transformations that could substantially alter the functional characteristics of animal health or behavioural and cognitive needs or functioning of research animals should be discussed on a case by case basis, and committee records should indicate that they have been discussed. A proper animal-ethics committee will include members who can and will articulate the rationale for concern about such experiments, even if they themselves may not endorse this rationale.

Committees should develop a forum in which the broad ethical acceptability of animal biotechnology is discussed. Such a forum may involve committee workshops and policy discussions, or general workshops and seminars made available to the entire faculty of researchers.

These are provisional suggestions that may need to be revised as committees (and society at large) learn more about animal biotechnology. For now, they provide a basis for research to continue while a great deal of philosophical uncertainty and confusion remains over the relationship between biotechnology and animal welfare.

Beyond the procedural research ethic for animal use, there are broader questions for the research agenda. Although the few studies that have been published show that animal biotechnology can be consistent with standards typically applied to the welfare of research animals, the small number of studies, the equivocal results and anecdotal reports of occasional welfare problems indicate that there is a need for continued and better research on the welfare of transgenic and cloned animals. Such research should be conducted under the still developing but now established paradigm of animal-welfare science. Studies will need to be conducted with some sophistication in order to avoid confounding factors. Given standard methodological and philosophical issues associated with assessment of animal welfare, it will continue to be possible to draw different conclusions from the same data and to aggregate results in ways that either favour or disfavour the comparison of biotechnology to more traditional programmes of livestock breeding. There should thus be a sustained effort within the scientific community to arrive at a consensus on which blend of standards and which forms of reporting data provide the most meaningful assessment of the welfare question. There should also be some effort to help members of the broader public, especially those interested in animal protection, evaluate highly aggregated or generalized statements about the link between biotechnology and animal welfare with a critical eye.

In closing with such a long list of 'should' statements and specific recommendations for animal-welfare committees, it may appear that the field of research ethics for animal biotechnology is left in a state of philosophical disarray. While it is obvious that many unsolved problems remain, the shift to a procedural-pragmatic approach that institutionalizes a process for combining seriousness of purpose, consideration for animals, philosophical pluralism and social learning should be regarded as a significant achievement. The problems with which I close are much more focused and amenable to amelioration (if not final solution) than are the grand puzzles of intrinsic value and the metaphysics of human–animal relations. I regard that as progress.

References

Appleby, M.C., 1998. Genetic engineering, welfare and accountability. *Journal of Applied Animal Welfare Science*, 1 (3), 255-173.

Appleby, M.C., 1999. *What should we do about animal welfare?* Blackwell Science, Oxford.

Balzer, P., Rippe, K.P. and Schaber, P., 2000. Two concepts of dignity for humans and non-human organisms in the context of genetic engineering. *Journal of Agricultural and Environmental Ethics,* 13 (1), 7-27.

Bekoff, M., 1994. Cognitive ethology and the treatment of non-human animals: how matters of mind inform matters of welfare. *Animal Welfare,* 3 (2), 75-96.

Bovenkerk, B., Brom, F.W.A. and Van den Bergh, B.J., 2002. Brave new birds: the use of animal integrity in animal ethics. *Hastings Center Report,* 32 (1), 16-22.

Bruce, D.M. and Bruce, A., 2000. Animal welfare and use. *In:* Hodges, J. and Han, I.K. eds. *Livestock, ethics and quality of life.* CABI Publishing, Wallingford, 53-78.

De Cock Buning, T., 2000. Genetic engineering: is there a moral issue? *In:* Balls, M., Van Zeller, A.-M. and Halder, M.E. eds. *Progress in the reduction, refinement and replacement of animal experimentation.* Elsevier Science, Amsterdam, 1457-1464. Developments in Animal and Veterinary Sciences no. 31.

Donnelley, S., 1990. Animals in science: the justification issue. *Hastings Center Report,* 20 (3 Suppl.), 8-13.

Durant, J., Bauer, M.W. and Gaskell, G. (eds.), 1998. *Biotechnology in the public sphere: a European sourcebook.* The Science Museum, London.

Fox, M.W., 1990. Transgenic animals: ethical and animal welfare concerns. *In:* Wheale, P. and McNally, R. eds. *The bio-revolution: cornucopia or pandora's box.* Pluto Press, London, 31-54.

Fraser, D., 1999. Animal ethics and animal welfare science: bridging the two cultures. *Applied Animal Behaviour Science,* 65 (3), 171-189.

Frewer, L.J., Howard, C. and Shepherd, R., 1997. Public concerns in the United Kingdom about general and specific applications of genetic engineering: risk, benefit and ethics. *Science, Technology and Human Values,* 22 (1), 98-124.

Heap, B.A. and Spencer, G.C.W., 2000. Animal biotechnology: convergence of science, law and ethics. *In:* Hodges, J. and Han, I.K. eds. *Livestock, ethics and quality of life.* CABI Publishing, Wallingford, 27-52.

Heitman, E., 2002. The human care and use of animals in research. *In:* Bulger, R.E., Heitman, E. and Reiser, S.J. eds. *The ethical dimensions of the biological and health sciences.* 2nd edn. Cambridge University Press, Cambridge, 183-191.

Hoban, T.J. and Kendall, P.A., 1993. *Consumer attitudes about food biotechnology.* North Carolina Cooperative Extension Service, Raleigh, NC.

Holland, A., 1995. Artificial lives: philosophical dimensions of farm animal biotechnology. *In:* Mepham, T.B., Tucker, G.A. and Wiseman, J. eds. *Issues in agricultural bioethics.* Nottingham University Press, Nottingham, 293-306.

Holland, A. and Johnson, A., 1998. *Animal biotechnology and ethics.* Chapman and Hall, London.

Hughes, B.O., Hughes, G.S., Waddington, D., et al., 1996. Behavioural comparison of transgenic and control sheep: movement order, behaviour on pasture and in covered pens. *Animal Science,* 63 (1), 91-101.

Jaenisch, R. and Wilmut, I., 2001. Don't clone humans! *Science,* 291 (5513), 2552.

Jennings, M. and Miller, J., 2000. Harmonizing IACUC practices. *In:* Balls, M., Van Zeller, A.-M. and Halder, M.E. eds. *Progress in the reduction, refinement and replacement of animal experimentation.* Elsevier Science, Amsterdam, 1705-1711. Developments in Animal and Veterinary Sciences no. 31.

Keulartz, J., Korthals, M., Schermer, M., et al., 2002. Ethics in a technological culture. *In:* Keulartz, J., Korthals, M., Schermer, M., et al. eds. *Pragmatist ethics for a technological culture*. Kluwer Academic Publishers, Dordrecht, 3-21. The International Library of Environmental, Agricultural and Food Ethics no. 3.

Kunkel, H.O., 2000. *Human issues in animal agriculture*. Texas A&M University Press, College Station.

Lanza, R.P., Cibelli, J.B., Faber, D., et al., 2001. Cloned cattle can be healthy and normal. *Science,* 294 (5548), 1893-1894.

Lekan, T., 2003. *Making morality: pragmatist reconstruction in ethical theory*. Vanderbilt University Press, Nashville.

Linzey, A., 1990. Human and animal slavery: a theological critique of genetic engineering. *In:* Wheale, P. and McNally, R. eds. *The bio-revolution: cornucopia or pandora's box*. Pluto Press, London, 175-188.

Marie, M., Edwards, S., Von Vorell, E., et al., 2003. Animal bioethics courses in Europe: state of the art, Paper presented at the IVth EurSafe Congress, Toulouse, 20-22 March.

McDonald, J.F. (ed.) 1991. *Agricultural biotechnology at the crossroads: biological, social and institutional concerns*. National Agricultural Biotechnology Council, Ithaca. NABC Report no. 3.

Mepham, T.B., Combes, R.D., Balls, M., et al., 1999. The use of transgenic animals in the European union: the report and recommendations of ECVAM workshop 28. *Alternatives to Laboratory Animals,* 27 (1 Suppl.), 21-43.

Mepham, T.B. and Crilly, R.E., 1999. Bioethical issues in the generation and use of transgenic farm animals. *Alternatives to Laboratory Animals,* 27 (1 Suppl.), 847-855.

Ministry of Agriculture, Fisheries and Food, 1993. *Report of the Committee on the Ethics of Genetic Modification and Food Use (the Polkinghorne Committee report)*. HMSO, London.

National Research Council NRC, 1996. *Guide for the care and use of laboratory animals*. National Academy Press, Washington DC. [http://www.nap.edu/readingroom/books/labrats/]

National Research Council NRC, 2002. *Animal biotechnology: science based concerns*. National Academy Press, Washington DC. [http://www.nap.edu/books/0309084393/html/]

Office of Laboratory Animal Welfare OLAW, 2002. *Institutional Animal Care and Use Committee guidebook*. 2nd edn. National Institutes of Health, Bethesda. [http://grants.nih.gov/grants/olaw/GuideBook.pdf]

Orlans, F.B., 1990. Policy issues in the use of animals in research, testing and experimentation. *Hastings Center Report,* 20 (3 Suppl.), 25-30.

Rifkin, J., 1995. Farm animals and the biotechnology revolution. *In:* Wheale, P. and McNally, R. eds. *Animal genetic engineering: of pigs, oncomice and men*. Pluto Press, London, 19-38.

Rollin, B.E., 1986. The Frankenstein thing. *In:* Evans, J.W. and Hollaender, A. eds. *Genetic engineering of animals: an agricultural perspective*. Plenum, New York, 285-298. Basic Life Sciences no. 37.

Rollin, B.E., 1989. *The unheeded cry: animal consciousness, animal pain and science*. Oxford University Press, Oxford.

Rollin, B.E., 1995. *The Frankenstein syndrome: ethical and social issues in the genetic engineering of animals*. Cambridge University Press, Cambridge.

Rowan, A.N., 1990. Ethical review and the animal care and use committee. *Hastings Center Report*, 20 (3 Suppl.), 19-24.

Russell, W.M.S. and Burch, R.L., 1959. *Principles of humane experimental technique*. Metheun, London.

Russow, L.M., 1999. Bioethics, animal research and ethical theory. *The ILAR Journal*, 40 (1), 15-21.

Ryder, R., 1995. Animal genetic engineering and human progress. *In:* Wheale, P. and McNally, R. eds. *Animal genetic engineering: of pigs, oncomice and men*, Vol. 1-16. Pluto Press, London.

Sandøe, P., Nielsen, B.L., Christensen, L.G., et al., 1999. Staying good while playing God: the ethics of breeding farm animals. *Animal Welfare*, 8 (4), 313-328.

Sparks, P., Shepherd, R. and Frewer, L.J., 1995. Assessing and structuring attitudes towards the use of gene technology in food production: the role of perceived ethical obligation. *Basic and Applied Social Psychology*, 16 (3), 267-285.

Thompson, P.B., 1997. Science policy and moral purity: the case of animal biotechnology. *Agriculture and Human Values*, 14 (1), 11-27.

Thompson, P.B., 1998. Report of the NABC ad-hoc committee on ethics. *Journal of Agricultural and Environmental Ethics*, 10 (2), 105-125.

Thompson, P.B., 1999. Ethical issues in livestock cloning. *Journal of Agricultural and Environmental Ethics*, 11 (3), 197-217.

Van Reenen, C.G. and Blokhuis, H.J., 1993. Investigating welfare of dairy calves involved in genetic modification: problems and perspectives. *Livestock Production Science*, 36 (1), 81-90.

Van Reenen, C.G. and Blokhuis, H.J., 1997. Evaluation of welfare of transgenic farm animals: lessons from a case study in cattle. *In:* Nilsson, A. ed. *Transgenic animals and food production: proceedings from an international workshop in Stockholm, May 1997*. Royal Swedish Academy of Agriculture and Forestry, Stockholm, 99-106.

Verhoog, H., 1992. The concept of intrinsic value and transgenic animals. *Journal of Agricultural and Environmental Ethics*, 5 (2), 147-160.

7b

Comments on Thompson: Research ethics for animal biotechnology

Mieke Boon[#]

Rules and principles of 'animal care and use' committees

In these comments I will compare the functioning of DECs (dierexperimentencommissies[1]) in The Netherlands, with the functioning of Animal Care and Use Committees (IACUC) in the United States as described by Paul Thompson[2]. Several aspects of the functioning of IACUCs in the Netherlands appear to be more or less similar to Thompson's description of the situation in the United States, but there also seem to be differences. As in the US, institutions in The Netherlands are required to constitute an internal DEC that will review and approve protocols. A DEC advises the licence holder on whether the protocol is acceptable from an ethical point of view. Contrary to the situation in the US, a governmental department ('veterinaire inspectie') is responsible for inspecting whether animal experiments are performed under licence, and whether animals used for research in the institution are in fact being used as indicated. This department also enacts rules concerning proper treatment of the animals. In my experience, an external governmental organization that sets certain rules and that controls whether rules and procedures are being followed contributes to the proper functioning of DECs because different roles are divided in this way. This provides the possibility of a more co-operative and open interaction between a DEC and researchers in the process of producing a protocol[3].

Besides governmental laws there may be some locally set rules or guidelines, which have been developed by precedents for similar cases. Examples of such rules are: "in decapitating animals anaesthesia should be used", "anaesthesia with an ether pot is not allowed", "solitary housing of rats should be avoided", "in applying this model for multiple sclerosis, measures should be taken for proper urinating of the animals". As can be seen, those rules aim at diminishing the amount of suffering. Usually these requirements lead to an adaptation of the proposal. Local guidelines and rules are not often laid down in an explicit code. The main reason for this may be that an explicit code often raises the impression that proposals are evaluated by a set of rigid rules, which as a consequence puts the burden of proof against procedures at the side of the committee.

Another class of local rules or guidelines consists of ethical principles that may lead to disapproval of an animal experiment, e.g., "animal experiments with a purely scientific aim are not allowed when causing major suffering", or "experiments with primates are not allowed at this university". With regard to very delicate issues, such as the use of primates, the board of a university can take the decision to lay down an

---

[#] Department of Philosophy, University of Twente, P.O. Box 217, 7500 AE Enschede, The Netherlands. E-mail: m.boon@utwente.nl

explicit rule. If not, those ethical principles play a role in the deliberation of a DEC, but may very well be overruled when stronger interests are at stake, such as 'important' scientific understanding, scientific reputation or Nobel prizes. Therefore, such principles do not function as ethical principles that judge what animal experiments are to be rejected irrespective of any interests.

Hence, like in the US ethical evaluation is not based on consistency with basic principles or norms with regard to the intrinsic value of an animal. From my point of view, a decision that (certain) animal experiments are no longer accepted, which is based on an ethical principle that is related to the intrinsic value of animals, should be a governmental decision and cannot be taken by a local animal-care and use committee. As long as society has not democratically decided to reject (certain) animal experiments principally, it would be unjustified for local DECs to take such decisions. As a consequence, decisions of DECs are utilitarian because they have to weigh different interests: besides care for animals, also interests of patients, researchers, 'science', university, consumers and companies have a part in the ethical considerations. But I agree with Thompson that this should not be thought of as involving a classical cost–benefit style of justification. Nevertheless, it is important to recognize the sketched practice as legitimate in order to understand that principle ethical difficulties with animal experiments will not be 'solved' by setting up animal use and care committees. In conclusion, decisions of DECs are not guided by general ethical principles concerning the intrinsic value of animals, since such principles would disapprove of (a certain class of) animal experiments irrespective of any other interests. On the contrary, decisions of DECs involve a utilitarian approach where interests of both animals and people are taken into account.

What should also worry us, and here I also agree with Thompson, is that nothing in this procedural approach itself guarantees that the right ethical questions regarding animal use are being considered. In my view, this problem occurs at two different levels. First, how a DEC is doing its job is not externally examined[4]. Thus, different DECs may uphold different standards. To a certain extent the nature of the approval reflects the culture and values of the institution, as Thompson puts it, but, as I see it, also those of 'society'. This brings me to the second level at which we should worry about whether the right ethical questions are considered. As I already pointed out, since DECs are held responsible for the ethical assessment of animal experiments, confusion may arise about who should decide on ethical principles concerning the intrinsic value of animals. As I stated earlier, principle decisions concerning the prohibition of (certain) animal experiments – i.e. legislation in which primacy is being given to the intrinsic value of animals instead of utilitarian arguments – ought to be taken democratically at a governmental level, not at the level of DECs at particular institutions. This latter point is often disregarded in critiques on the functioning of ethical committees. I will get back to this point below.

The role of ethical expertise

A consequence of this utilitarian approach by DECs is that in most cases proposed animal experiments are approved when animal suffering is reduced 'as much as reasonably possible'. For, proposed animal experiments usually aim at 'accepted' interests. Diminishing the amount of animal suffering helps to balance the interests of different parties, and thus meets utilitarian requirements. As a consequence, many discussions in the committee deal with rather technical or scientific questions concerning details of how animal welfare in the experimental set-up can be improved and how the amount of animals used can be reduced. According to Thompson judging animal welfare also involves an ethical decision because such judgments presuppose ideas about how to interpret 'animal welfare'. In his view, this would require specific ethical expertise in the committee. Although this may be the case at the level of methodology in animal-welfare research, in my view this cannot be a topic for an animal-care and use committee. My disagreement with Thompson on this point may have to do with the fact that in The Netherlands veterinary and animal-welfare expertise is represented by a specialist in service of the institutions, who advises both DEC and researchers. Usually, the committee accepts his or her assessment of animal suffering. Therefore, assessment of animal welfare in an experimental set-up is a technical or scientific – and not an ethical – question in the context of the committee.

One could ask, as Thompson did, why ethical expertise is required in animal-care and use committees. The committee's major goal appears to be meeting the aims of reduction, replacement and refinement ('the three R's'), and weighing the interests of different parties involved. This is an ethically set goal, but approaching this goal does not seem to require specific ethical expertise. Here, I assume that specific 'ethical expertise' is only required when deciding on whether a certain new ethical principles is at stake. Aiming at the reduction of animal suffering and weighing interests of different parties, requires a discussion between prudent people endowed with specific scientific expertise, practical intelligence and *common morality*, not specific *ethical expertise*. Nevertheless, ethical problems concerning fundamental issues may be at stake every once in a while, and ethical expertise may be required to recognize these issues and direct the discussion to a specific ethical level. Examples are whether animal experiments on smoking or drugs addiction are ethically justifiable, or how to deal with research paid by an industry that is commercially dependent on the results, or how to judge the new research possibilities that are generated by using transgenic or knock-out animals. And also the issue raised by Thompson about whether animal welfare should be reduced to the prevention of stress and pain occasionally comes up in discussions of the committee. Despite the importance of these ethical issues, they often cannot productively be dealt with within the context of the committee or institution. For in most cases, as soon as scientific and/or societal significance is recognized, ethical principles concerning the intrinsic value of animals appear to be too weak to 'outweigh' these interests. In such cases a strong implicit argument seems to be that rejecting a protocol for ethical principles concerning the intrinsic value of animals would be unfair because such a decision will put a researcher or a research group at a disadvantage in its community[5]. In my view, these problematic issues can – and should – only be productively discussed at the public or governmental level. Possibly, those members of the committee endowed with specific ethical expertise, could take the responsibility to bring those ethical issues to the level of the public debate. These debates may finally lead to new laws, such as the law that animal experiments for cosmetic products are prohibited.

The use of transgenic and knock-out animals

In order to discuss problems of animal welfare concerning the use of transgenic and cloned animals in experiments, Thompson makes use of insight about the use of such animals in agriculture. In my view this approach is of minor value because several problems apparently relevant to biotechnology in agriculture are not relevant to animal experiments and, on the other hand, important problems in animal experiments are overlooked. Before pointing out specific problems in animal experiments, I will explain in what respect the use of biotechnology differs in the two fields.

Firstly, contrary to agriculture where transgenic animals are used, so-called 'knock-out' animals are often used in animal experiments. Transgenic animals carry a foreign gene, for instance a transgenic cow that produces a certain pharmaceutical product in her milk. In knock-out animals one or more genes are 'knocked-out' in order to inhibit or destroy certain physiological functions. Knock-out animals are used to study the role of those genes in physiological processes. Secondly, Thompson discusses the problems of animal welfare that may arise as a consequence of manipulation of embryos and in-vitro fertilization. As far as I know, this technique is not used in breeding knock-out or transgenic animals (mostly mice) for experiments.

Beside ethical problems with genetic modification of animals as such, several ethical problems that may not be relevant to agriculture do occur however in applying biotechnology to experimental animals. (1) Changing the animal's DNA can have various effects on the animal's functioning. Usually those effects cannot be predicted. Therefore, in animal experiments consequences to animal welfare should be assessed for each specific case. As I already stressed, this assessment is largely dependent of the expertise of animal-welfare specialists. (2) A specific problem with the breeding of knock-out animals is the number of 'surplus' animals, i.e. animals that do not have the right genotype. From Mendelian calculations one can understand why, in order to produce one pure knock-out, at least 12 animals are wasted, and in order to produce a double knock-out, this number increases exponentially. Since homozygotic knock-out animals are often infertile after several generations, one can imagine that enormous numbers of surplus animals are a consequence of experiments with 'knock-out' animals. (3) The production of knock-out animals seems to provide scientists with almost infinite possibilities to do research. This may easily lead to a significant increase of animal experiments. (4) From a methodological point of view one may doubt whether results with knock-out animals can always be translated to human diseases. As we know by now the genetic make-up often appears to be very adaptive and is able to compensate for imperfections. Moreover, in different people a particular disease may be caused by different combinations of several genes. Therefore, one can doubt whether experiments with knock-out animals will usually lead to the sophisticated results biotechnology seems to promise.

Taking these problems together it becomes obvious that possibilities of biotechnology in animal experiments burden us with new and serious ethical problems, which to some extent are different to problems of biotechnology in agriculture. Therefore, the suggestions of Thompson for a proper approach of these problems by IACUCs seem to be of limited value. In my view, the assessment of animal experiments with transgenic or knock-out animals does not in principal differ from other assessments[6]. For reasons I have already pointed out, new ethical problems concerning the use of transgenic or knock-out animals should be recognized by DEC members but cannot be properly dealt with by the committee. Again, those ethical problems should be treated in public and governmental debates. There, members of animal-care and use committees can contribute from their specific expertise.

From these examples that help to illustrate how DECs actually function, it becomes obvious that DECs have a delicate role in the dynamics between animal experimentation, public opinion and legislation. For some it may be disappointing that animal-care and use committees do not act as moral crusaders in expelling animal experiments. I have tried to make clear that applying ethical principles that overrule utilitarian considerations is unjustified if those ethical principles are not accepted by

Parliament. In other words, institutions such as a DEC should only overrule utilitarian considerations by ethical principles if such ethical principles are enacted in laws. Nevertheless, members of animal-care and use committees could play an important role in showing the dilemmas and possible ethical viewpoints, which may lead to new laws that are shaped by those principles.

---

[1] In the Netherlands the abbreviation DEC stands for animal experimentation committee, not for animal ethics committee

[2] Comments on this paper are based on my experience as the chairman of a Dutch Animal Care and Use Committee of the Free University in Amsterdam for the last six years

[3] In this context it should be recognized that many researchers genuinely care about their animals. Often the societal relevance of their research is required by themselves too. These researchers appreciate input of the DEC in meeting the three R's (reduction of animal use, refinement to reduce animal suffering, replacement of animal experiments by other methods)

[4] In The Netherlands a long-running debate on the public accountability of DEC's concerns this issue

[5] Exceptions are proposals where it is generally accepted that the societal benefit is in no proportion to sincere suffering of the animals. However, due to self-censorship of researchers and subsidizing institutions, these situations appear to be very rare

[6] Even the argument that the amount of suffering caused by the genetic transformation cannot be predicted properly does not differ in principle from difficulties in predicting the suffering caused by other treatments

ETHICS FOR LIFE SCIENTISTS AS A CHALLENGE FOR ETHICS

8a

How common morality relates to business and the professions

Bernard Gert[#]

Common morality

I am aware that many people believe there is no substantial agreement on moral matters. I am also aware that there is even less agreement on the adequacy of any account of morality. I believe that these views are due to the understandable, but mistaken, concentration on controversial moral issues concerning the environment, especially the treatment of animals, and such issues as abortion and euthanasia. Some people, particularly philosophers, do not realize that such controversial matters form only a very small part of those matters on which people make moral decisions and judgments. Indeed, most moral matters are so uncontroversial that people do not even make any conscious decision concerning them. The uncontroversial nature of these matters is shown by everyone's lack of hesitancy in making negative moral judgments about those who harm others simply because they do not like them. It is shown by the same lack of hesitancy in making moral judgments condemning unjustified deception, breaking of promises, cheating, disobeying the law and not doing one's duty.

An explicit, precise and comprehensive account of morality should help to make clear the uncontroversial nature of most moral decisions and judgments, whether personal, professional or business. Such an account should also help in understanding some of the controversial moral problems that arise in the practice of business and the professions such as engineering, law, medicine and science. Common morality provides a framework on which all of the disputing parties can agree, making clear what is responsible for the disagreements and what might be done to manage that disagreement. An account of common morality is an account of the moral system that is already implicitly used by people when dealing with everyday moral problems. Nothing new is being proposed and there should be nothing surprising in the account that is presented. The point of providing an explicit account of the moral system is to enable it to be used by people when they are confronted with new, difficult or controversial moral decisions[1].

Those who deny the possibility of an explicit, precise and comprehensive account of morality may actually be denying that any systematic account of morality provides a unique answer to every moral problem. It is, in fact, true that the common moral system does not provide a unique solution to every moral problem. Not every moral problem will have a single best solution, that is, one that all equally informed impartial rational persons would prefer to every other solution. In the overwhelming numbers of instances, common morality does provide a unique answer, however most of these cases are not interesting. Only in very few situations does an explicit account

---

[#] Department of Philosophy, Dartmouth College, Hanover, New Hampshire 03755, USA. E-mail: bernard.gert@dartmouth.edu

of morality settle what initially seemed to be a controversial matter, e.g., whether there was a moral difference between a rational refusal of food and fluids by a competent terminally ill patient and that patient's refusal of medical treatment. However, even in the more usual situation where controversial cases do not have a unique answer, common morality is still often quite useful. It places significant limits on legitimate moral disagreement, i.e., it always provides a method for distinguishing between morally acceptable answers and morally unacceptable answers and provides the basis for a fruitful and respectful discussion of the issue. It promotes moral tolerance and makes clear that attempts to reach a consensus do not require compromising one's moral integrity.

The fact that legitimate moral disagreement on some issues is compatible with complete agreement on many other issues seems to be almost universally overlooked. Many philosophers seem to hold that if equally informed impartial rational persons can disagree on some moral matters, they can disagree on all of them. Thus many philosophers hold either that there is no unique right answer to any moral question or that there is a unique right answer to every moral question. The unexciting, but correct, view is that although most moral questions have unique right answers, some do not. Although moral agreement is far more common than moral disagreement, since moral disagreement is far more interesting to discuss and hence discussed far more often, people wrongly conclude that moral disagreement is more common than moral agreement.

Morality as an Informal Public System

The existence of a common morality is shown by the widespread agreement on most moral matters. Everyone agrees that such actions as killing, causing pain or disability and depriving of freedom or pleasure are immoral unless one has an adequate justification for doing them. Similarly, everyone agrees that deceiving, breaking a promise, cheating, breaking the law and neglecting one's duties also need justification in order not to be immoral. No one has any real doubts about these claims. People do disagree about the scope of morality, e.g., whether embryos, foetuses and non-human animals are impartially protected, or protected at all, by morality, however everyone agrees that actual moral agents, i.e., those whose actions are themselves subject to moral judgment, are impartially protected. Doubt about whether killing an animal or an embryos needs to be justified, does not lead to any doubt that killing a moral agent needs justification. Similarly, people disagree about what counts as an adequate moral justification for some particular act of killing or deceiving and on some features of an adequate justification, but everyone agrees that what counts as an adequate justification for one person must be an adequate justification for anyone else in the same situation, i.e., when all of the morally relevant features of the two situations are the same. This is what is meant by saying that the moral rules must be obeyed impartiality.

Morality is a public system, and like all public systems it has the following two characteristics. (1) All persons to whom it applies, i.e., those whose behaviour is to be guided and judged by that system, understand it, i.e., know what kinds of behaviour the system prohibits, requires, discourages, encourages and allows. (2) It is not irrational for any of these persons to accept being guided and judged by that system. The clearest example of a public system is a game such as football. This game has an inherent goal and a set of rules that form a system that is understood by all of the players; and it is not irrational for all players to use the goal and the rules of the game to guide their own behaviour and to judge the behaviour of other players by them. Although a game is a public system, it applies only to those playing the game. Morality is a public system that applies to all moral agents; all people are subject to morality simply by virtue of being rational persons who are responsible for their actions. This may explain Kant's claim that the demands of morality are categorical, not hypothetical.

One of the tasks of a moral theory is to explain why sometimes, even when there is complete agreement on the facts, genuine moral disagreement cannot be eliminated, but the theory must also explain why all moral disagreement has legitimate limits. It is very easy, as noted above, to overlook that unresolvable moral disagreement on some important issues, e.g., abortion, is compatible with total agreement in the overwhelming number of situations in which moral decisions are made, or the overwhelming number of cases on which moral judgments are made. This agreement is based on agreement about the nature of morality, that it is a public system with the goal of reducing the amount of harm suffered by those protected by it. Everyone agrees that morality prohibits some kinds of actions (e.g., killing and breaking promises) and encourages certain kinds of actions (e.g., preventing the suffering of pain). But it is acknowledged that it is sometimes morally justified to do a prohibited kind of action even when it does not conflict with another prohibition, but rather when it conflicts with what is morally encouraged. Breaking a trivial promise in order to aid an injured person is regarded by all as morally acceptable.

In addition to the disagreement about who is in the group impartially protected by morality, another important source of irresolvable disagreement in moral decisions and judgments is due to differences in the rankings of harms, including differences in how one ranks different probabilities of harms. Disagreements about the proper speed limit is a disagreement about whether the certain deprivation of some small freedom to millions, viz., the freedom to drive between 100 kilometres an hour and 130 kilometres an hour, is justified by the probability that there will be fewer people killed and injured in automobile accidents. Indeed, many political disagreements are about this kind of difference in rankings. They can be regarded as a conflict between the rankings of freedom and welfare, e.g., how strict the regulations concerning building construction or pollution should be. Even the disagreement about the scope of morality can be seen as a conflict between freedom, the freedom of moral agents to treat non-moral agents as they want, and welfare, the welfare of these non-moral agents, e.g., non-human animals. The presence of these kinds of irresolvable moral disagreements must be reflected in an adequate account of morality.

Morality, like all informal public systems, presupposes overwhelming agreement on most matters that are likely to arise. However, unlike formal public systems, it has no established procedures or authorities that can resolve every moral disagreement. There is no equivalent in morality to the referees or umpires in professional sports. When there is no unique right answer within morality and a decision has to be made, the decision is often made in an *ad hoc* fashion, e.g., people may ask a friend for advice. If the moral disagreement is on some important social issue, e.g., abortion, the problem is dealt with by the political or legal system. Since abortion and the treatment of animals are irresolvable moral questions, and it is necessary that the society have some rules about these matters, the question is transferred to the legal and political system. They resolve the questions on a practical level, but they do not resolve the moral questions, as is shown by the continuing intense moral debate on these matters.

Failure to appreciate that morality is an informal public system has caused considerable confusion when talking about public policies, not only with regard to abortion and the treatment of animals, but also in many other areas, such as pollution control. It is assumed that if morality does not directly provide a solution to the problem, it can always provide an indirect solution by means of an appropriate voting procedure. It is sometime mistakenly said that a just solution, which must be a morally acceptable solution, is one that is arrived at by a democratic voting procedure. The justness or moral acceptability of a solution to a problem cannot be determined by any voting procedure, for a majority can vote to deprive members of a minority group unjustifiably of some freedom. The moral acceptability of a solution is determined by the moral system; but a democratic voting procedure does provide a morally acceptable way of choosing between morally acceptable alternatives. This democratic voting procedure may be the morally best way to determine which morally acceptable alternative will be adopted, but it does not make that solution either morally acceptable or the morally best solution.

Common morality, which is the moral system that all thoughtful people use, though usually not consciously, presupposes that people are vulnerable and fallible. Its goal is to lessen the amount of harm suffered by those protected by it. These harms: death, pain, disability, loss of freedom and loss of pleasure, are ranked differently by different people, but everyone regards all them as harms, i.e., as things that all rational persons would avoid unless they had an adequate reason not to. Although all competent persons must know the language of their society, that is, how to use it in understanding what others are saying and in making themselves understood by others, they need not be able to articulate the rules they are following. Similarly, although all moral agents must know the kinds of actions that morality prohibits, requires, discourages, encourages and allows, they need not be able to articulate the rules they are following.

The moral system

Common morality recognizes both the vulnerability and the fallibility of people. It includes (1) rules prohibiting acting, or attempting to act, in ways that cause, or significantly increase the probability of causing, any of the five harms that all rational persons want to avoid, and (2) ideals encouraging the prevention of any of these harms. It also includes (3) a two-step procedure for deciding when it is justified to violate a moral rule. This procedure reduces the likelihood of making mistaken decisions and judgments due to failure to consider all the relevant facts and the biases toward self and friends that most people have. The first step of this procedure requires describing the violation solely by means of its morally relevant features; the second step requires estimating the consequences of everyone knowing that such a violation is allowed and consequences of everyone knowing that it is not allowed.

It is useful to provide an explicit, precise and comprehensive account of the justified moral system that is common morality. It is not useful, but dangerous, to provide a system that can be applied mechanically to arrive at the correct solution to a moral problem; not all moral problems have unique correct solutions. Common morality only provides a framework for dealing with moral problems in a way that is acceptable to all moral agents insofar as they are informed, impartial and rational. This justified moral system does not provide a unique right answer to every moral question. All impartial rational persons accept common morality as a public system that applies to all moral agents, but it does not eliminate all moral disagreement. In what follows I shall attempt to make explicit the details of the common moral system. I do not think that anyone will find anything surprising in this explication.

The moral rules

The first five moral rules prohibit directly causing the five harms that all rational persons would avoid unless they had an adequate reason not to avoid them.
Do not kill. (Includes prohibition on causing permanent loss of consciousness)
Do not cause pain. (Includes prohibition on causing mental pain, e.g., anxiety)
Do not disable. (Or, do not cause loss of physical, mental or volitional abilities)
Do not deprive of freedom. (Includes freedom from being acted upon as well as depriving of resources and opportunity to act)
Do not deprive of pleasure. (Includes depriving of sources of pleasure)
The second five moral rules include those rules which, when not followed, in particular cases usually, but not always, cause harm and which always result in harm being suffered when they are not generally followed.
Do not deceive. (Includes more than lying)
Keep your promises. (Equivalent to Do not break your promise)
Do not cheat. (Paradigm involves violating rules of a game)
Obey the law. (Equivalent to Do not break the law)
Do your duty. (Equivalent to Do not neglect your duty)
The term 'duty' is being used in its everyday sense to refer to what is required by one's role in society, primarily one's job, not as philosophers customarily use it, which is to say, simply as a synonym for "what one morally ought to do". Using the term 'duty' in its ordinary sense allows for a clear discussion of the relationship between the common moral system and professional ethics and business ethics. A proper understanding of this relationship requires getting clear about the duties of professionals.

The moral ideals

The moral ideals include such precepts as "Prevent (postpone) death", "Prevent pain", "Relieve pain", "Prevent disabilities", "Prevent the loss of freedom". Whereas the moral rules provide limits to what any person is allowed to do regardless of his aims or goals, the moral ideals set forth aims or goals that all persons are encouraged to adopt. Whereas it is possible to obey the moral rules all of the time impartially with regard to everyone, it is impossible to act on the moral ideals either all of the time or impartially with regard to everyone. I can be obeying the moral rules, that is, not violating them, even when I am sleeping, but I cannot be following the moral ideals when I am sleeping. The moral rules are all stated as prohibitions, or can, without any change in meaning, be stated as prohibitions. That is why I cannot only obey them all of the time, I can also obey them impartially with regard to everyone. If I obey the rule, "Do not kill" I am obeying with regard to everyone, even people I do not even know anything about. The moral ideals set forward positive goals, so that in following them I always follow them with regard to particular persons or kinds of persons. It is not possible to prevent or relieve pain for everyone.

This difference between the moral rules and the moral ideals explains why it is appropriate to enforce obedience to the moral rules. We can make a person liable to punishment for all serious unjustified violations of a moral rule because no one is ever supposed to violate any moral rule unjustifiably. It makes no sense to make a person liable to punishment for failing to follow a moral ideal, for then all of us would almost always be liable to punishment. As long as they are not breaking any moral rule, everyone is encouraged to follow the moral ideals toward any group of persons that they want, as often as they want. A person who does not follow any moral ideals at any time, is not a morally good person, but he need not be an immoral person. It is a serious mistake to think of morality as consisting solely of prohibitions and requirements. The moral ideals, the following of which is encouraged, not required, are such a significant part of morality, that they can sometimes justify violating a moral rule.

Justifying violations of the moral rules

The major value of simple slogans like the Golden Rule, "Do unto others as you would have them do unto you" and Kant's Categorical Imperative, "Act only on that maxim that you could will to be a universal law" are as devices to persuade people to act impartially when they are contemplating violating a moral rule. However, given that these slogans are often misleading, a better way to achieve impartiality is to consider whether it is rational to favour that violation even if everyone knows that this kind of violation (or a violation in these circumstances) is allowed. An impartial rational person would not favour everyone knowing that a kind of act is allowed if the consequences of everyone knowing that this kind of act being allowed would have worse consequences than everyone knowing that this kind of act is not allowed. Only this willingness that everyone know that this kind of violation is allowed, what I call publicly allowing the violation, shows that one is not making special exceptions for oneself. A willingness to publicly allow the violation is necessary for impartiality; if a person violates a moral rule when he would not publicly allow that kind of violation, he is acting arrogantly.

What counts as the same kind of violation, or the same circumstances, is determined by the morally relevant features of the situation. Two violations count as the same kind of violation or a violation in the same circumstances if all of the following questions have the same answers: The answers to these questions are the morally relevant features of the act.

1. Which moral rule is being violated?
2. Which harms are being caused, avoided (not caused), and prevented?
3. What are the relevant beliefs and desires of the person harmed?
4. Does one of the parties have special duties toward the other?
5. Which goods are being gained?
6. Is a violation of a moral rule being prevented?
7. Is a violation of a moral rule being punished?
8. Are there alternative actions that are morally better?
9. Is the action being done intentionally, or only knowingly?
10. Is the situation an emergency?

Once the correct description of the kind of violation or the circumstances of the violation is provided, then it is necessary to estimate the consequences of that kind of violation being publicly allowed and of it not being publicly allowed. If a person favours this kind of violation when a friend does it to a stranger, he must also favour this kind of violation when a stranger does it to a friend. A person who favours one violation must favour all violations of the same kind. But two different people may agree about the kind of violation in question, but still disagree about whether they favour that kind of violation. They can also sometimes disagree about whether there is even a violation of a moral rule, that is, they can disagree about the interpretation of a moral rule. This disagreement, like the disagreement about whether to favour a violation of a moral rule, can sometimes be settled by estimating the consequences of publicly adopting different interpretations. If everyone agrees that the consequences of adopting one interpretation is better than adopting any other, then that is the interpretation that should be adopted. However, sometimes not all informed impartial rational persons will agree and then the difference in interpretation cannot be resolved. Unresolvable disagreements among impartial rational persons arise because of:

a) different views about the scope of morality, e.g., whether it provides impartial protection, some protection, or no protection to nonhuman animals or to foetuses;
b) different rankings of the harms and benefits involved;
c) different ideologies, such as different unverifiable beliefs about human nature and society, i.e., different beliefs about how people would behave if they knew that a certain kind of act were allowed;
d) different interpretations of a moral rule, e.g., whether the act counts as deceiving or killing. (As noted above, this difference is based on the three previous differences.)

Relations between common morality, the duties of professionals and the duties of those of business

Professional ethics is not distinct from common morality. It does not provide exemptions from common morality, rather, a profession takes on certain duties that are not duties for those not in that profession. However, all of these duties must be compatible with the framework provided by common morality. No profession can have a duty to do what all informed, impartial, rational persons would consider morally unacceptable. Nor can any business impose duties on its employees that would be considered morally unacceptable by all informed, impartial, rational persons. A job cannot impose duties that are morally unacceptable. A driver of a getaway car does not have a duty to help bank robbers escape after a bank robbery, even if he has been paid to do so. An employee of an advertising company does not have a duty to help compose an advertisement that will entice young people into smoking or taking any other addictive drug. Scientists who are employed by a cigarette company do not have a duty to help make that product more addictive.

In this context, it is important that the term 'duty' not be interpreted as philosophers normally interpret it, namely as a synonym for 'what one is moral required to do'. Duties arise from a social role or some special circumstances. Judges, parents, teachers, doctors and nurses have duties that arise from their jobs and professions. People do not have a duty not to kill although they are morally required not to kill unless they have an adequate justification for doing so. The same is true of all of the first nine moral rules, we do not have a duty to obey them, but we are morally required to obey them unless we have an adequate justification for not doing so. Unless the term 'duty' is restricted to the moral requirement imposed by a person's social role, it will be impossible to provide a clear account of the relationship between common morality and professional duties and the duties of those in business[2].

To claim that no one can have a duty to do what is morally unacceptable is not to deny that having a duty to do something may make what would otherwise be morally unacceptable into an action that is morally acceptable. It is normally morally unacceptable for one person to kill another person, or deprive him of his freedom, when doing so is not necessary to prevent his causing an even greater harm. But if the person is an employee of the government, then he may have a duty to kill or deprive of freedom someone who has committed and been legally found guilty of a crime for which death or deprivation of freedom is the prescribed penalty. Some might claim that any killing that is not necessary to prevent more killing, e.g., killing in self-defence, is never morally acceptable, and so no one can have a duty to kill a person as a legal punishment. This is a plausible view, but it is a controversial one, and some people defend the justifiability of the death penalty. When it is not clear that an action is morally unacceptable, a person can still have a duty to do it. Further, everyone agrees that a person can have a duty to deprive someone of his freedom even when that is not necessary to prevent his causing greater harm. Some system of punishment involving deprivation of freedom is accepted by all, so it is accepted by all that a person can have a duty to deprive people of their freedom, without even considering whether that particular act of freedom deprivation is necessary to prevent greater harm.

Duties related to employment
Scientists normally have duties that are related to their employment. It is not clear that scientists qua scientists have any duties. If a scientist is not employed by a private or public employer but is simply doing science for his own enjoyment, he has no duties specifically related to his profession as a scientist. However, if he works at a university, or some other non-profit research organization or some business, he may have duties that require him to conduct his research in certain areas, to publish that research in reputable journals, and perhaps to aid in the application of his research to practical problems. He is, of course, morally required not to misrepresent his research findings, but that is also true of the scientist who is working on his own. The prohibition not to misrepresent one's research is not some special duty related to scientists, but simply an application of the general moral requirement not to deceive. The interesting question is whether scientists who are employed by a firm or organization may have a duty not to reveal their research findings to people outside of the firm or organization for which they work. Normally, there is no moral requirement for a scientist to report his research, so if it is part of the job description for a scientist not to disclose his research results without the approval of the company, he may have a duty not to reveal his research findings. However, in special cases there may be a moral requirement to report. If the firm has published data, e.g., about a drug or about the effect of some project on the environment, which the scientist has discovered are false, the scientist may have a duty to report his findings. Similarly, if a drug company hires a scientist to do research on a drug that they are selling or plan to sell, and the scientist finds out that the drug is either not as effective as other drugs that are cheaper and as easily available, or that it has serious harmful side effects, that scientist may be morally required to report his findings to some outside party, even if the company has imposed a job requirement not to report.

How can it be that, if an independent researcher has no moral requirement to report any of his findings, someone who supposedly has a job requirement not to reveal, may have a duty to report his findings? Ironically, it may be because he is part of a company that is deceiving the public, that he is morally required to report. Not reporting false data within the scientist's area of special research may be taken as his being part of a conspiracy to deceive. Scientists at the various tobacco companies who had discovered the facts, may have acted immorally when they did not report their findings that cigarettes were both addictive and causing cancer. However, if they had a moral requirement to report, it was not because scientists qua scientists have a duty to report all findings that might be helpful to the general public, but because they worked for a company that was deceiving the public and not to report was to be a party to that deception.

There may be organizations of scientists, as there are of engineers, in which those who are practicing scientists of one kind or another, e.g., biologists, chemists, geologists etc., take on special duties simply in virtue of becoming that kind of scientist. However, I do not know what special duties these could be except in relationship to employment. Once employed, scientists may have the standard duties that employees have, plus other that are related to their special expertise. There may be duties not to participate in any fraudulent activities by the company, but these duties are otiose, since everyone is morally required not to participate in fraudulent activities. What may make it not entirely redundant to have such a duty is that the duty may require reporting to outside authorities that the company is putting out data that, on the basis of his own research, the scientist knows to be false. This duty to report does not come into play until after the scientist has informed the company of this research and given them the opportunity to correct their previous false claims. It is only when the company persists in putting out the false information that the scientist has a duty to disclose his research to those outside the company.

Everyone is morally required not to deceive, but that is not the same as being morally required to tell the truth. Between lying and telling the truth, is staying silent. Doctors not only are morally required not to lie to their patients, they have a duty to provide information to their patients about their diagnoses and prognoses. To withhold that information counts as deceiving. That is what having a duty to disclose does, it adds to the moral requirement of not lying, the moral requirement not to remain silent, that is, not to withhold information that the patient is entitled to have. Of course, sometimes deception, lying or withholding information is justified. If the consequences of telling the truth are serious enough, i.e. the patient is at a serious risk of committing suicide, a doctor may be justified in withholding and, if necessary, even lying. But absent, these serious consequences, a doctor is not only morally required not to lie, he is also morally required not to withhold.

Similarly, scientists, like everyone else, are morally required not to deceive, but they are not normally required to tell all that they know on the basis of their scientific research. Scientists are normally allowed to remain silent. However, it is quite plausible to maintain that they cannot remain silent when the organization that they are working for is putting out as information, what they know to be false. It may be that scientists have a duty not to allow deception by the organization for which they work, at least with regard to information that they have in their capacity as a scientific researcher. This would mean, paradoxically, that scientists are morally required to report on misinformation put out by their own companies or organizations, but not morally required to report on misinformation put out by other companies or organizations.

What I have said about scientists may also apply to accountants and lawyers who are employed by a company. They cannot remain silent if the company is doing something that is contrary to the standards that their profession is committed to and with regard to which they have special expertise. Accountants cannot remain silent when companies are reporting their finances in a fraudulent way. Lawyers cannot remain silent when companies are clearly violating the law. What exactly they are required to do, may be determined by the duties that people take on when they enter that profession. But part of being a professional involves having duties that go beyond what people are normally morally required to do. They must not only not break the moral rules themselves, they have a duty not to allow the organization for which they work to break the rules in an area for which they have a professional responsibility. This is the special added duty that they have, one that is clearly not incompatible with what is morally required of everyone, but rather goes beyond what is morally required for others.

Those in business who are not professionals, have the same moral requirements not to deceive as anyone else, but they do not have the same requirement to expose the deception of the company or organization. If an automobile company advertises their cars in a way that a salesman for that automobile believes to be false, he is not morally required to tell potential customers that the advertisement is false. He should not use the false information himself in order to sell the car, but his is not morally required to correct the customer, or report to some outside source, that he believes the information to be false. The salesman has no special expertise that makes him have a duty not to remain silent when he believes information that the company has put out is false. But, of course, it would be acting on a moral ideal to reveal that the company is putting out false information, that is, it would be morally good to do it. However, the cost to the person may be great enough that he is not prepared to act on this moral ideal to prevent deception. A normal employee who lacks the courage to report immoral activities by his company, is not doing a morally good action, but that is not the same as acting immorally.

Professionals do have a duty not to remain silent when they believe information that is within their area of expertise that the company has put out is false. A professional has a duty to reveal deception by the company or organization when the deception is within his area of expertise. If he does not reveal that deception, he is complicit with it. That is the consequence of his having the duty to reveal the deception. For the professional to reveal that information, does not count as acting on a moral ideal, rather he is obeying the moral rule requiring him to do his duty. It is morally required. A scientist who lacks the courage to report immoral activities by his company, is not simply failing to do a morally good action, he is acting immorally. Indeed, many of the duties of professionals require actions that would count as acting on moral ideals if they were not duties. This reinforces the point that I have been emphasizing, that the duties of professionals are never incompatible with common morality, but rather are additions to it. It is an important point about duties that they can never be incompatible with what is required by common morality.

But duties of professionals are not limited to actions that would be following moral ideals by persons without such duties. A judge has a duty to decide cases that come before her impartially. She is not allowed to shape her decisions in order to benefit friends. Contrary to what is commonly claimed, impartiality is not a moral ideal. It is not better to give impartially to every charity that requests a donation rather than limiting one's donations to one or two charities and give them much larger amounts. Indeed, one can even choose the charities to which one contributes by purely arbitrary means, e.g., a person can give to a charity concerned with heart disease because her father died of a heart attack. Impartiality is required when considering violations of a moral rule, and by certain duties, but not otherwise. As John Stuart Mill says, "Impartiality, however, does not seem to be regarded as a duty in itself, but rather as instrumental to some other duty; for it is admitted that favour and preference are not always censurable, and indeed the cases in which they are condemned are rather the exception than the rule. A person would be more likely to be blamed than applauded for giving his family and friends no superiority in good offices over strangers, when he could do so without violating any other duty". (Utilitarianism, Chapter 5, paragraph 10.)

It is not merely with regard to deception by his company or organization that a scientist or other professional has a duty to do something. Any time that his company or organization in unjustifiably violating a moral rule when that is within his area of expertise, a professional has a duty to speak out. If his company is polluting a water supply, a geologist working for that company must act to stop that pollution. The same is true of any other kind of pollution, when that pollution is within the scientist's area of expertise. Since the scientist has a duty to do something, failure to act is to be in collusion with the pollution. This is in contrast with a non-professional employee of the company. For such an employee it would be following a moral ideal to do something to stop the pollution, but except in special circumstances, a regular employee of a company does not have a duty to do something about the pollution. He is not in collusion if he fails to do something, even though he is not acting in a morally good way[3]. Of course, the president of the company, or that official who is responsible for the pollution, is acting immorally, but that is because, like everyone, he is morally required not to harm people unjustifiably.

---

[1] Gert, B., 1998. *Morality: its nature and justification.* Oxford University Press, New York contains a more extended account of morality, and of the moral theory that justifies it. A shorter account of morality and its justification is contained in *Common Morality*, (forthcoming)

[2] I am ignoring the duties that arise from special circumstances, e.g., being in a unique position to help someone avoid a serious harm without any significant cost to oneself, because these kinds of duties are not relevant to the topic of the relationship between common morality and business and professional ethics

[3] See Hennessey, J.W. and Gert, B., 1985. Moral rules and moral ideals: a useful distinction in business and professional practice. Journal of Business Ethics, 4 (2), 105-115. German translation reprinted in *Wirtschaft und Ethik*, ed. by Hans Lenk and Matthias Maring, 1992

8b

Comments on Gert: Gert's common morality: old-fashioned or untimely?

Jozef Keulartz[#]

According to Gregor McLennan, "We are all pluralists now". During the last decades a "glacial shift away from monism, towards pluralism" has occurred. "Where once the onus was on pluralists to bounce off, and try to dismantle, the grand monistic edifices, today any credible 'big picture', will have to be very careful not to appear to obliterate or devalue perceived plurality" (1995, p. 99). Late modernity or postmodernity is characterized by pluralism in science and society, in theory and in practice. We are on the road from unity to diversity, to difference, decentring, dissemination, deconstruction and discontinuity, to mention the main terms that circulate to indicate the ongoing process of pluralization.[1]

Ethics too has taken the road from unity to diversity, albeit with some reluctance it seems to me. An early example of a strong form of pluralism can be found in Christopher Stone's *Earth and Other Ethics: The Case for Moral Pluralism* (1987). According to Stone, the problem with contemporary ethical theories is that they are still "aiming to produce, and to defend against all rivals, a single coherent and complete set of principles capable of governing all moral quandaries" (p. 116). This monism becomes problematic as soon as all kinds of exotic entities enter the moral arena, such as future generations, the dead, embryos, animals, the spatially remote, tribes, trees, robots, mountains and art works. As a result of their emergence on the moral scene the assumptions that unify ordinary morals are called into question.

A less extreme and more moderate pluralism can be found in Beauchamp and Childress' well-know *Principles of Biomedical Ethics* (1994). They call their theory, which finds its source in the common morality and uses principles as their structural basis, a pluralistic theory. The formulation of their famous four principles is inspired by different and even diverging theories. This is rather seen as an advantage than as a disadvantage. "We stand to learn from all of these theories. Where one theory is weak in accounting for some part of the moral life, another is often strong. Although each type of theory clashes at some point with deep moral convictions, each also articulates norms that we are reluctant to relinquish ... We reject the assumption that one must defend a single type of theory that is solely principle-based, virtue-based, rights-based, case-based, and so forth. In moral reasoning we often blend appeals to principles, rules, rights, virtues, passions, analogies, paradigms, parables, and interpretations. To assign priority to one of these factors as the key ingredient is a dubious project" (Beauchamp and Childress 1994, p. 111).

The new intellectual situation, characterized by a marked anti-foundationalism in epistemology as well as in ethics, has evoked new questions and doubts. Doesn't pluralism itself constitute just another type of doctrine, a new grand narrative and a

---

[#] Applied Philosophy Group, Wageningen University, P.O. Box 8130, 6700 EW Wageningen, The Netherlands. E-mail: Jozef.Keulartz@wur.nl

new monism – with a pluralist face? Will a politics of difference, which indiscriminately and without any exception gives equal attention to every voice that makes itself heard, not gradually blur into a politics of indifference? Is it possible to impose limits on the proliferation of differences, behind which a viable pluralism changes into an infertile relativism, an unsound eclecticism or a dangerous nihilism?

In this intellectual climate Bernard Gert, together with Charles Culver and Danner Clouser, has argued for 'A Return to Fundamentals', as the subtitle of their book on Bioethics from 1997 says. In 1990 Clouser and Gert launched a fierce attack on 'principlism', a term they use to designate all theories composed of a plural body of potentially conflicting *prima facie* principles. These theories, including the theory of Beauchamp and Childress, "fail to provide a unified theory of justification or a general theory that ties the principles together as a systematic, coherent, and comprehensive body of guidelines, with the consequence that the alleged action-guides are ad hoc constructions lacking systematic order" (1994, p. 100).

In the present intellectual climate, Gert's plea for a return to fundamentals and to a unified theory sounds surprising and absolutely refreshing. The question remains, however, whether this form of neo-foundationalism (as it is sometimes called) is untimely (in the sense of Nietzsche's *Unzeitgemässe Betrachtungen*) or just old-fashioned?

Gert maintains that there is only one morality, common morality, and that it is possible to provide an explicit description of this common morality that is clear, coherent and comprehensive. This description of the moral system can be achieved by transforming the mostly implicit 'know how' of competent moral agents into explicit 'know that', analogous to the way linguists provide an explicit description of the grammatical system by systematizing the utterances of competent speakers. After describing the moral system, Gert goes on to explain and justify its nature by relating it to the universal features of human nature such as fallibility, vulnerability and rationality.

Although Gert's approach certainly provides us with many interesting and useful insights (see Matthias Kettner 2003), it is in my view not without its difficulties and problems. Even if we assume that there is only one moral reality, one moral universe, it is highly unlikely that that there is only one unique description of this universe possible. I agree with Richard Rorty that "there are many descriptions of the world and of ourselves possible, and the most important distinction is that between those descriptions which are less and those which are more useful with respect to a specific purpose" (1999, p. 27). Or to quote the famous German philosopher Wilhelm Dilthey: "The pure light of truth can be seen by us only in variously broken rays" (Dilthey 1931, p. 222). No single description is capable of capturing reality in its full versatility, but is one-sided by necessity. In my opinion, this also applies to Gert's description.

Overall, the moral theory that is offered by Gert can be understood as a form of rule consequentialism. What is or is not morally acceptable is largely dependent on moral rules and justifiable violations of them, whilst the criteria for moral acceptability are given in terms of the avoidance of harm. Thus the theory is a rule-based one with the locus of moral value centring on the avoidance of harm. One significant feature of the theory that distinguishes it from other consequentialist theories is the importance of publicity in cases where a violation of a moral rule is being considered.

Gert distinguishes ten moral rules, neatly divided into two distinct categories, the first five rules prohibiting *directly* causing all of the basic harms (death, pain, disability, deprivation of freedom and of pleasure) and the second five prohibiting those kinds of actions that *indirectly* cause these same harms (do not deceive, keep your promises, do not cheat, obey the law and do your duty). As Beauchamp and Childress (1994) correctly state: "Clouser and Gert rely almost exclusively on nonmaleficence in their ethical theory" (p. 318).

To me, it seems highly implausible that the exploration and explication of the rich moral life could result in the conclusion that the agreement that exists among our moral judgments "is based on agreement about the nature of morality, that it is a public system with the goal of reducing the amount of harm suffered by those protected by it". That we all agree about the nature of morality seems questionable and that we all agree that its goal is to reduce the amount of harm suffered seems even more difficult.

The one-sidedness of Gert's account of common morality is evident from the fact that it can do no more justice to considered moral judgments than all other consequentialist theories can. The most important of these judgments are judgments about distributive justice and judgments about respect for autonomy. I believe there is a connection here: the absence of notions such as justice, fairness and equality in the final analysis points to a disregard for the concept of autonomy or dignity of individual men and women. I will not go into this connection but restrict myself to questions of distributive justice.

I agree with Michael Walzer that a careful analysis of our moral practices shows that there exists a plurality of distributive principles relative to different social goods or sets of goods, like free exchange, desert and need.

That Gert too cannot avoid using such principles, albeit implicitly, comes to light if we, for instance, examine his interpretation of job discrimination against qualified people of a particular race, religion or ethnic background from his latest book on *Common Morality* (forthcoming). According to Gert this kind of discrimination counts as a violation of the rule 'Do not deprive of freedom'. However, as Gert notes, "it does not normally count as depriving a person of an opportunity if another more qualified person is hired for the job". But this judgment is only valid on the basis of a hidden assumption, in the form of the following criterion of fairness or distributive justice: one must give equal consideration to every qualified candidate, and one must take into account only relevant qualities.[2]

That Gert's interpretation of job discrimination is really problematic also comes to the fore if we compare this kind of negative discrimination with positive discrimination. In the first case men and women have been discriminated against in the distribution of jobs, because of their membership of an ethnic or religious group, and not for any reason having to do with their individual qualifications. In the second case it is argued, for the sake of fairness and redress, we should now discriminate in their favour, even set aside a certain number of offices exclusively for them. Now, Gert lacks the conceptual means to distinguish between positive and negative discrimination because the amount of harm will be the same in both cases. And this is a problem because this distinction is reflected in our moral judgments: whereas (nearly) everybody will reject negative discrimination as totally unjustified, the opinion on positive discrimination stands divided, which means that this policy must be considered (in Gert's own words) 'weakly justified'.[3]

To close my comment, I will briefly go into the notion of 'sustainable development', which increasingly can be found in mission statements and professional codes, due to the recent emergence of 'corporate social responsibility' (or 'corporate citizenship'). According to the well-known definition of sustainable development from the Brundtland-report, "sustainable development is development that meets the needs of the present generation without compromising the ability of future generations to meet their own needs". In fact, this definition combines justice *within* the current generation with justice *between* the current generation and future generations. In other words, it combines intra-generational justice with inter-generational justice.

The definition of sustainable development makes reference to 'needs'. In Walzer's theory need is the central criterion of distribution in the sphere of security and welfare. Walzer assumes "that every political community must attend to the needs of its members as they collectively understand those needs; that the goods that are distributed must be distributed in proportion to need; and that the distribution must recognize and uphold the underlying equality of membership" (1983, p. 84). According to Walzer, every distribution in proportion to need is inevitably always also a redistribution, the strongest shoulders carrying the heaviest load. The problem with Gert's account of morality is that all acts of redistribution as well as all acts of redress, due to their 'idealistic' character, appear as supererogatory actions, that is, as actions that no one can expect anyone to perform. This is, I believe, what makes his account too minimalist to function as a reliable public guide for the behaviour of all moral agents, business and professionals included.

References

Beauchamp, T.L. and Childress, J.F., 1994. *Principles of biomedical ethics*. 4th edn. Oxford University Press, New York.
Clouser, K.D. and Gert, B., 1990. A critique of principlism. *Journal of Medicine and Philosophy*, 15 (2), 219-236.
Dilthey, W., 1931. *Weltanschauungslehre: Abhandlungen zur Philosophie der Philosophie*. Teubner, Leipzig. Gesammelte Schriften no. 8.
Gert, B., forthcoming. *Common Morality*.
Gert, B., Culver, C.M. and Clouser, K.D., 1997. *Bioethics: a return to fundamentals*. 2nd edn. Oxford University Press, New York.
Kettner, M., 2003. Die Konzeption der Bioethik von Bernard Gert, Charles M. Culver und K. Danner Clouser. *In:* Düwell, M. and Steigleder, K. eds. *Bioethik: eine Einführung*. Suhrkamp, Frankfurt, 88-104.
MacLennan, G., 1995. *Pluralism*. Open University Press, Buckingham.
Rorty, R., 1999. *Philosophy and social hope*. Penguin Books, London.
Stone, C.D., 1987. *Earth and other ethics: the case for moral pluralism*. Harper & Row, New York.
Walzer, M., 1983. *Spheres of justice: a defence of pluralism and equality*. Basic Books, Princeton.

---

[1] In philosophy, the culminating point of the road from unity to diversity was reached in the work of Jean-François Lyotard, who with much brouhaha announced the end of all grand stories ('meta-écrits'). Instead of the one grand narrative there is a multitude of narratives, language games, discourse genres or vocabularies.
[2] There are of course exceptions to this general rule: "For many offices, only minimal qualification is required; a large number of applicants can do the work perfectly well, and no additional training would enable them to do it better. Here fairness seems to require that the offices be distributed among qualified candidates on 'first come, first served' basis (or through a lottery)" (Walzer 1983, p. 135-136).

[3] Another example concerns Gert's interpretation of the rule 'Do not cheat'. "Successful cheating results in the cheater gaining some advantage over others participating in that activity". It is not at all clear how this would lead to more harm than sheer bad luck or a strong opponent who can easily win without having to cheat. In short, this is a moral violation only if some notion of fairness is used.

9a

Research as a challenge for ethical reflection

Marcus Düwell[#]

Introduction

Research in the life sciences has different types of far-reaching impacts on the lives of human beings. Those impacts are partly intended but partly not foreseen when the research got started. Obviously, the relevant moral aspects of such developments go far beyond the moral responsibilities of individual researchers. It seems, however, a task for different actors in society to reflect on the moral dimensions of research and new technologies. But there is hardly any agreement on the question whether such an ethical evaluation of research is possible and in which way it should be done. We are faced with a plurality of moral convictions, a diversity of ethical theories and an increasing variety of technical, ecological, economic and social aspects constructing the context of modern research and the conditions of their application. In order to assess those possibilities I will have a look at the normative framework that can be found in our generally expected practice of ethical examination. I especially want to ask whether the instruments available are sufficient for an adequate moral evaluation of the new research developments. In that context it seems a crucial point of discussion to what extent the methodology of such applied ethics is able to deal with the insecurity of the future developments of research and the unclear options of future applications. Some scientific activities, for instance, are not directly harming individual rights but are ambivalent in their possible applications. The only thing we can expect is that these activities will obviously entail far-reaching consequences for our lives. Since we cannot foresee them, a moral evaluation of those scientific and technological enterprises is difficult. This difficulty is not only a (more or less) technical problem of the technology assessment, but it is also a question for the methodology and the theoretical framework of research ethics to what extent they are able to take those dimensions of research and new technologies into account. In this respect I want to examine the available normative frameworks, whether they are able to achieve an ethical evaluation of research that is able to take the social and political impacts of those scientific developments into account; an ethical evaluation that is appropriate to the complexity and importance of research and new technologies for modern society.

Normative Framework

When the World Medical Association in Helsinki and Tokyo accepted the rights of the patient to decide at free will and well-informed about the treatment he or she would undergo, an important step was set for the protection of the individual. It was a

---

[#] Faculty of Philosophy, Ethics Institute, University of Utrecht, Heidelberglaan 8, 3584 CS Utrecht, The Netherlands. E-mail: marcus.duewell@phil.uu.nl

milestone in the transformation of the conviction and the self-understanding of medicine – a discipline the history of which was always accompanied by moral reflection. High moral expectations had been carried towards the profession of a physician, but at the same time the attitude of the physicians against the patient had been a very ambiguous one. Being in need of medical treatment, human beings were often in danger of becoming depersonalized by the knowledge of physicians, who saw themselves often as the administrators of the well-being of the patient. The high moral impact of the self-image of medicine has been one of the reasons for a latent paternalistic attitude in the medical ethos.

Nowadays, the notion of a patient's autonomy and self-determination has become intrinsically linked with our view of the medical ethos. Perhaps we are no longer really able to appreciate the importance of that change for the moral orientation, which took place in the 1960s and 1970s (see for the historical development: Jonsen 1998). If we look at the moral convictions to be found in the international declarations and conventions, we will encounter a central position of human rights. And mostly the content of human rights can be explained primarily by the right of the individual to decide freely about the treatment that he or she would undergo. If one examines the practice of ethical committees that evaluate experiments involving human beings, one can see that, in general, the central aspects of the evaluation include the expected result of the experiment, the risk for the subject involved and the protection of his or her decision to be taken at free will and well-informed. Moreover, the European Convention on Human Rights and Biomedicine considers the protection of the free decision of the individual as a core right to be protected. This position is even central to the whole structure of the convention, and correspondingly a great part of the convention deals with the question how to treat people who are not able to give consent (Council of Europe 1997, Art. 6, 7, 17, 20). In short, if the free and informed consent starts to become the cornerstone of the moral conviction, it becomes most important to challenge cases where the patient cannot give consent and where therefore the informed consent is not an option to protect the patient's interests. We are thus faced with a moral framework that puts great emphasis on the question how to secure the free decision of the individual against tendencies in medical practice to overrule the free will of the person. Taking into account the history of some physicians in the Nazi concentration camps during World War II, we cannot help of being glad about such a development. The same development is to be found if we look at the central importance of human rights in the secular moral convictions of Western societies. But it is the question whether such a normative framework of protecting the individual choice is sufficient for the challenge that ethical reflection has to face with respect to the new developments in the life sciences.

If we consider the most influential book on bioethics from the last decades, "Principles of Biomedical Ethics" by Tom Beauchamp and James Childress (2001), of which a revised edition is published every few years, we could get the feeling that the analysis I have offered is a bit too hasty. According to Beauchamp and Childress, autonomy is only one of four principles. Besides autonomy, beneficence, non-maleficence and justice are also part of that biomedical ethos of which the principles – according to the authors – are used by all participants in discussions about medical ethics, irrespective of the theoretical presuppositions they make. Although it falls outside my scope to give more detailed comments on this approach (cf. Clouser and Gert 1990), I would like to argue that the set of normative forces would not become richer via such additional notions. Beauchamp and Childress use the notion of non-maleficence to emphasize the need to protect the individual from suffering direct harm. With the notion of beneficence they introduce an internal teleology of the acting of physicians. With the notion of justice they refer in the first place to aspects of equality in the treatment of patients and the general accessibility of the health-care system. But we can interpret the general approach of this set of *prima facie* principles as an expression of the protection of individual rights against inadequate treatment by physicians. In general their reflections remain within the context of the medical practice and try to secure the patient in that context. In fact, we can interpret it as the articulation of that ethos of autonomy and self-determination I have mentioned before. The long-term perspective of the impact of the life sciences on the life of human beings is not of central importance to them.

Viewed more generally, it seems to me that the compatibility of the right to self-determination with a contractualist perspective may explain its general acceptance in biomedical ethics. In a secular world it seems necessary for moral norms to be compatible with an ethos of the self-interested perspective of each individual to be accepted. Only those norms will be successful that are not asking for individuals with good moral intentions and that can (at least in the long run) be seen as an adequate interpretation of the interests we all have. This ethos is interested in fundamental security, which can be offered by a secular morality in the form of a social contract. Let me emphasize that I am not defending such a position, but that I am interested in understanding why it is successful.

If this assessment of the chances of moral reflections in a secular world is right, we have to interpret the central position of the ethos of free and informed consent as an articulation of a set of moral convictions concentrated around the idea that we all want a situation of general security, which can only be guaranteed in a society where the individual can be sure that his will is accepted and where the governmental institutions are in the first place legitimized by their ability and task to protect the security of the individual. In the concept of Thomas Hobbes, we are confronted with the idea that those institutions have to protect us against destructive tendencies of our anthropological constitution. In later discussions, contractualists have tried to become independent of such theoretical demanding and controversial anthropological presuppositions. In order to defend – with John Rawls – the priority of the right against the different notions of the good, it is sufficient to presume that we shall not from our very nature act in a peaceful way and that consensual solutions will not appear without specific regulations and institutions. To legitimate the need for moral regulations, we do not have to presuppose any bad intentions of the human being; we do not need a 'negative' anthropology, but it is sufficient to refer to the fact that moral conflicts are not avoidable without conscious decisions.

A moral protection of our right to self-determination can be interpreted from such a fundamental contractualist idea. It will limit the scope of a generally accepted morality to only those moral norms that are compatible with the negative rights of every agent, and it will assume that the rights that we concede to each other are strictly mutual. In most Western societies, this ethos of the protection of our negative freedom has of course been enriched with some ideas of a welfare state and of a government that acts supportively for its people. But this additional, positive or supportive ethos has an unclear position in the bioethical discussion. In the political debate concerning bioethics it disappears very easily. If the public bio-political discourse has the task to produce a minimal consensus, it is the normal procedure that only the protection of the right of self-determination of the individual will be consensual. Thus, only the negative right to self-determination will stay as a strong normative approach, and everything beyond that minimal ethos will be a matter of choice between the conflicting convictions. One good example is, for instance, the European discourse about the treatment of human embryos. We are often confronted with the observation that there is a conflict between the European countries (whereby generally my native country Germany is mentioned explicitly in that context)[1]. Most of the people who look at these political processes with a realistic view, articulate the expectation that there will be no consent to be reached in the next future. The consequence is that tolerance for the position of the other is claimed. To avoid misunderstandings, I have to add that I do not hold the opinion that a human embryo has the same moral status as a person does, even though I believe that the embryo is to be protected in several respects (Düwell 2003). But the point that is important to me here is a theoretical one. If we ask for tolerance where it regards conflicting positions, we already assume that the point in question is answered, because if we ask for acceptance of a position we presuppose that that position is in principle morally acceptable. But, if the pro-life position is right, the destruction of a human embryo is a violation of human dignity and defending such an action cannot be acceptable at all. To my mind, the example clarifies how the creation of a minimal consensus is a way to reduce the possible moral convictions in the discourse to a more or less contractualist moral position.

More examples are found in bioethical theories. Tristram Engelhardt's *Foundations of Bioethics* (1996) for instance, illustrates the defence of the principle of autonomy and the principle of beneficence. Engelhardt attempts to explain a secular ethos that is to be understood as a minimal ethos. Very briefly stated, we are all members of groups that, although being different in several respects, share several moral convictions. We therefore have *moral friends* with whom we share a set of common values. For Engelhardt himself, the moral friends are to be found in some orthodox-Christian groups. But outside those groups of moral friends we are *moral strangers*. The minimal ethos is to be considered an explanation of that set of moral convictions that is necessary for the coexistence between moral strangers. Engelhardt explains that the ethos of acceptance of an individual's autonomy is a strong claim, even between moral strangers. If we are looking for a moral authorization of our action, we are looking for something that is not compatible with violence. The types of moral authorization may be different, we may have very different kinds of moral arguments, but looking for a moral foundation for our action always means that we do not want to solve our conflicts purely by the law of the jungle. In the centre of that minimal notion of morality we find the prohibition of 'unconsented-to force against the innocent'. This means that even between moral strangers that law is valid. And if we act against it, we shall be seen as a kind of outlaw, we are no longer members of a moral community at all. We can doubt whether Engelhardt is very successful in his reasoning why we are obliged to accept all norms that are to be regarded as a result of that minimal ethos that exists between moral strangers. Since he has no philosophical concept of a moral obligation, we can doubt whether that project is successful at all (A lucid critics is to be found in: Steigleder 2003; 1992). But Engelhardt goes further and claims that also a principle of beneficence has to be accepted within the minimal notion of morality. Even if we accept it, we can doubt whether a principle of beneficence has the possible impact of such a minimal ethos. In the logic of an ethos that is a kind of peace-making project between different communities of moral

---

[1] Of course it has to be mentioned that the disagreement within the European Societies is in general much bigger than the international disagreement.

believers, one can only argue for a cease-fire project, which is realized by a minimal ethos of autonomy.

The liberal concept and its critics

I do not assume that a more or less liberal, minimal ethos is the consciously chosen theory, which is dominant in the politically influential bioethical discourse. I would rather be inclined to a direction in which the ethos of the biopolitical discourse can be best understood by a contractualist point of view. I have not said very much about that hypothesis until now, but I would like to refer to some central presuppositions in the concrete ethical debate, which I believe have far-reaching consequences for the structure of the discourse. I want to emphasize this point of view in two ways, firstly a more theoretical and secondly a more practical one. In doing so, I want to outline some desiderata of the ethical discussion.

To start with the theoretical aspects: In a pluralistic society we are used to thinking in terms of a fundamental difference between the right and the plural ideas of the good. There is a plurality of moral convictions and most of the more or less liberal concepts of moral and political philosophy want to defend the position that there are some fundamental ideas of justice and human rights that should have priority. We can interpret the right as an overlapping consensus between different ideas of the good, we can interpret it as a set of basic convictions that are implications of the idea of a human person, or we can see it as a sort of minimal consensus between all those different moral approaches. Against this contractualist or liberal perspective criticism from different approaches has been articulated. The conservative *communitarians* are afraid that reducing morality to such a liberal, minimal ethos will destroy the moral energies, which are found in our moral communities (MacIntyre 1981). Those communitarians are afraid that on the one hand the chances for a philosophical foundation of a liberal morality are not very good and that on the other hand the liberal criticism against the traditional moralities will destroy the only sources which are available for morality at all. The very influential *ethics of care* stresses the point that a broad variety of relevant moral aspects is ignored if we reduce the centre of moral convictions to the idea of the right. According to these critics, the attitude of caring for the other has a priority in relation to the formulation of individual rights, and the scope of morally relevant aspects cannot be restricted to those claims which are implications of the protection of the liberty of the free and rational being. Between communitarianism and care ethics we find a variety of other approaches that are critical against that liberal ethos. We can mention the critics of authors like Amartya Sen or Martha Nussbaum, who argue against the liberal reduced perspective by stating that the moral framework has to be described in a way that is different from the way the liberal perspective describes it. Nussbaum attempts to legitimize moral claims in a way that the basic needs and basic capacities for all human beings have to be protected (Nussbaum 1990; 2000). She can argue in that way for a hierarchy of basic goods which are forming the foundation of moral evaluations. In doing so, she defends the modern idea of universality and the idea that there is a difference between a set of moral convictions, which is strictly obligatory for everyone, and a plurality of moral convictions where a diversity of values can exist and has to be accepted. But she does not restrict the scope of a binding morality to the protection of the negative rights.

Another kind of criticism against the liberal concept is found in several attempts to reintroduce virtue ethics, perfectionist concepts or concepts of care in the debate. The mentioned discussion criticizes the reduction of the liberal ethos to a concept of a person who is a strategically thinking, self-interested and atomistic individual. According to this criticism, we have to reintroduce emotions, motivations and meaning in the ethical discourse in order to enrich our moral universe. We have to deal with real persons and include their level of intentions and moral lives in our moral considerations, instead of drawing the 'veil of ignorance' over all concrete elements of life in the ethical discourse. The relation between those strategies and the liberal project is ambivalent. Some ethicists really want to present an alternative approach to the liberal society in an ethics of 'Lebenskunst' or in a new virtue ethics. Others see themselves in the framework of a pluralistic society and develop creative and stimulating moments in a liberal and pluralistic society. In that case the normative basic convictions of an ethos of autonomy, of free and informed consent are not touched by those reflections.

In such criticism of the liberal idea of a priority of the negative right for the different ideas of the good, a problematic confrontation will emerge. It seems that the liberal idea of rights is restricted to negative rights, which find a foundation in the liberal idea of a social contract. On the other side, we have a broad variety of moral convictions that are articulated against the liberal perspective, but which do not seem to have the pretension to be argued as morally right. My suggestion would be to criticize that alternative. Whether or not the scope of rights is to be identical with the negative liberal rights, is a matter of discussion. Here the role of moral philosophy in the concrete ethical debate becomes obvious. We have to ask why we are obliged to respect the rights of each individual. In doing so the question has to be put out in the open: what are the contents of the rights we are obliged to respect? The legitimization and the content of the moral rights will thus have some connection.

Ethical discourse and the complexities of new technologies

The second consideration will be on a more practical level. If the ethical discussion concentrates on the protection of the self-determination of every individual, it has to be asked on what aspects such a moral debate will be focussed. It seems that the ethical debate is in this way restricted to those moral aspects that arise when a technology is on its way into practice. The question in that kind of moral debate will be: how can we protect the individual against possible harm through the application of a technology? But, the development of technologies itself seems to be morally neutral. To my mind, the key restriction of the bioethical discourse is the ignoring of central aspects of the social dimensions of those technologies. And that has to do with the theoretical problems I have mentioned above. Let me explain that by describing some issues.

If one describes moral questions that are connected with the research of a human being, one can start with either an analysis of the process of the research, with the targets that are the aims of that research, or with the expected outcome. If the ethical evaluation starts with the research process, one will be concentrating on the methods that are used and on questions that have to do with the responsibility of the researcher. One will furthermore examine from a moral perspective whether or not specific rights and values of human beings are touched; perhaps the protection of animals will be taken into consideration as well. Here it is possible to describe the relevant aspects, because the circumstances are known and in general the responsibilities are known as well. If, on the other hand, one is more focused on the possible outcome of the research, one is faced with many more insecurities and unknown aspects. The range of research activities that are nowadays carried out under the label 'genomics', include a broad variety of research in biology, pharmacy, agriculture, veterinary medicine, human medicine and so on. It is not a research activity with a common methodology and a clear target. It is an ensemble of research activities, each with totally different application conditions. We only know that the expectations concerning the output are high. We expect that several fields in the life sciences can be changed, but we do not know precisely what the possibilities for application will be. We do not know what will change in medicine, pharmacy or veterinary medicine. We can also expect that some fields in practice will change although we did not have that in mind when the research got started. Since we do not have enough knowledge about the possible applications, we cannot know what will change in the different fields. If there are possibilities for creating pharmaceutical products that will cure a specific genetic disease, what will that mean for the whole idea of treatment in our clinics, for the perception of illness, for the self-understanding of patients and for the financing structure of the health-care system? What economic possibilities for agriculture will be connected with that research? In what way will our general concepts of illness, nature and bodily identity be changed? If the impact is so far-reaching, then we can expect that the consequences for several dimensions of life will be enormous.

For an adequate ethical framework for the evaluation of such scientific developments we have to wonder about the possible levels of moral regulation. In the framework of the ethics of an informed consent, we can first of all assure ourselves that no rights of human beings are touched or violated in the research process. Furthermore, we must ensure that in the application of the result of the research adequate control mechanisms are institutionalized to avoid harm to the people. Besides that, an ethics of free and informed consent, a liberal ethics, could fulfil its task in ensuring the possibility of autonomous decisions in dealing with the applications of such research. All of us, in our role as patients, consumers or citizens, have to be given the possibility to decide freely about the use of such developments. This means that measures should be taken to ensure that we are able to reject the use of such an outcome of genomics research if we want to. Furthermore, we should also consider the implementation of relevant measures in order to bring high-quality information to the public. If people shall decide freely, they have to be made competent to do so. This means that appropriate information materials have to be made available to ensure that competent decisions can be taken, and that infrastructures that give the consumer and citizen adequate access to all knowledge and information necessary for his free decision, have to be established.

What I have described here very briefly is the normal procedure in dealing with new scientific developments. It seems that the ethical approach has the advantage that it is not paternalistic. It is a liberal approach, in so far that it respects the different decisions of the citizen and the consumer and that it is compatible with different worldviews. It has the advantage of being clear concerning the responsibilities of the different actors in the field. The government has to take measures to support research, avoid harm and enable free decision-making. The consumer has to inform him- or herself, if he or she wants to decide freely. The scientists have to perform their job in a methodically correct manner, and have to make all scientific information available. Such an ethos is not burdened with high and idealistic expectations concerning the morality of different actors. The researcher only has to do a good job and be honest, that's enough.

All ethical reflections that go further are either a criticism of the fundments of a liberal society or deliberative considerations within the liberal framework. Reflections of the first kind are easy to criticize; reflections of the second kind are deliberative considerations inside the liberal society. The critical attitude towards the liberal society was articulated, for example, in the debate around communitarianism, where the priority of the right and the plurality of the conceptions of the good were no longer accepted. That debate was able to articulate moral aspects in the evaluation of new developments only in the framework of specific moral convictions of particular groups in society. Therefore, they were either paternalistic in forcing their moral convictions to society or they were unable to deal with the plurality of moral concepts in another than a traditionalistic way. A convincing alternative to the liberal project cannot be found in that way.

Concluding remarks

My sketch of the ethical landscape is aimed at explaining the problems for an ethical discourse that has to deal with the complex impact of new research. If ethics has the task to ensure autonomous decision-making, it will allow research to continue but will reduce the task of ethics to ensure free decision-making when dealing with the result of the research done. The moral evaluation of the question as to what impact that research has on our life will not take place. But, in fact, I think that, on the one hand, the task of ethical reflection goes further than ensuring autonomous decision-making, and, on the other hand, that it has to be seen within the framework of a liberal society. The question for the ethical debate should rather be what implications the protection of the rights of the individual and the respect we owe to each other has for new scientific developments. Is it enough to protect the possibility of free decision-making inside a room of alternatives and options that are already determined by the scientific development? In that concept, the alternatives between which the citizen can choose freely are already created by the scientific community. The structure of the room of decision-making will then not be the object of an ethical debate. The subject will find himself in a situation of decision-making where the options have already been structured in a way that the pathways of his choices are foreseeable. The question for an ethical reflection which wants to evaluate the scientific development, including the whole range of implications for society, economy and our private lives, thus has to be in what way an evaluation of research activities with their possible implications for our lives is possible, in order to avoid that moral reflection can only happen in a situation when the range of options and alternatives has already been decided by others.

In the framework of a liberal society I do not see an alternative to the exceptions of the priority of the respect we owe to each other for the different concepts of the good each of us can follow. But the question is what the content of those moral respects is. It is possible that we have to conceptualize it in such a way that research has to develop in its own inner logic and we have only to protect individuals against harm in the process of research as such. But it is also possible that our moral rights include much more. It is possible that the protection of the basic capacities of all has to be directed towards the hierarchy of the goods that should be protected. The range of moral rights has not to be restricted to negative rights, meaning that not only those measures are necessary that protect everyone against direct interference in the freedom of his acting. It is possible that there also are positive rights. This means that we owe to each other the support that we need in order to be able to live a good life. All the mentioned possibilities to interpret the content of the rights and obligations we have towards each other would have different impact on the moral evaluation of the impact of scientific activities in a liberal society. The complexity of the impact that research has on our society and existence forces us to open the discussion about the normative framework of such an evaluation. Reducing that normative framework to an ethos of free and informed consent does not enable us to deal with the complexity of the new developments.

References

Beauchamp, T.L. and Childres, J.F., 2001. *Principle of biomedical ethics.* 5th edn. Oxford University Press, New York.
Clouser, K.D. and Gert, B., 1990. A critique of principlism. *Journal of Medicine and Philosophy,* 15 (2), 219-236.
Council of Europe, 1997. *The convention on human rights and biomedicine.* [http://conventions.coe.int/treaty/en/treaties/html/164.htm]
Düwell, M., 2003. Der moralische Status von Embryonen und Feten. *In:* Düwell, M. and Steigleder, K. eds. *Bioethik: eine Einführung.* Suhrkamp, Frankfurt, 221-229.
Engelhardt Jr., H.T., 1996. *The foundations of bioethics.* 2nd edn. Oxford University Press, New York.
Jonsen, A.R., 1998. *The birth of bioethics.* Oxford University Press, New York.
MacIntyre, A., 1981. *After virtue: a study in moral theory.* Duckworth, London.
Nussbaum, M., 1990. Aristotelian social democracy. *In:* Douglass, R.B., Mara, G.M. and Richardson, H.S. eds. *Liberalism and the good.* Routledge, New York, 203-252.
Nussbaum, M.C., 2000. *Women and human development: the capabilities approach.* Cambride University Press, Cambridge.
Steigleder, K., 1992. *Die Begründung des moralischen Sollens: Studien zur Möglichkeit einer normativen Ethik.* Attempto, Tübingen.
Steigleder, K., 2003. Bioethik als Singular und als Plural: die Theorien von H. Tristram Engelhardt, Jr. *In:* Düwell, M. and Steigleder, K. eds. *Bioethik: eine Einführung.* Suhrkamp, Frankfurt, 72-87.

9b

Comments on Düwell: Research as a challenge for ethical reflection

Akke van der Zijpp[#]

The ethical question addressed in this paper concerns the moral responsibility of the researcher. This question is directly connected with the task of different actors in society to reflect on the moral dimensions of research and their interrelationships in terms of moral obligations and concerns. This aspect is not discussed in the paper.

When the physicians accepted the rights of the patient to informed consent, the ethos of autonomy and self-determination of the patient or individual human being became central in our thinking. The central position of autonomy is not, in Düwell's opinion, challenged by the four principles of Beauchamp and Childress. They include autonomy, beneficence, maleficence and justice, but when challenged the remaining and dominant position is human autonomy and self-determination.

The case of Herman, the first bull genetically engineered for production of (human) lactoferrin in cow's milk as human medicine and baby food

In 1990 the Advisory Committee on Ethics and Biotechnology in Animals of the Dutch minister of Agriculture, Nature Conservation and Fisheries designed a model for ethical evaluation. The model was based on the four principles of Beauchamp and Childress and added the principles of redress (precaution) and verification (democratic control). The Committee came to this evaluation accepting a zoocentric approach, inspired by philosophical concepts of Singer, Regan and Rollin. This report speaks about welfare and health of animals in particular as practical indicators for the four principles of Beauchamp and Childress. A few years later new animal-welfare laws in The Netherlands were more explicit in accepting integrity and intrinsic value of animals as basic concepts. Obviously animals cannot speak for themselves, but humans have the moral obligation to speak for them.

The ethical evaluation itself then takes the form of reviewing factual data, assessment of consequences for nature, the intrinsic value of animals, human health and welfare and of the environment, and weighing of values in terms of threat and violation of interests. For the above case of Herman the result was initially a conditional yes, with reporting about health and welfare being a central focus in the ongoing experiment regarding both transgenic and normal offspring of Herman. Later the decision was made to castrate the bull Herman to prevent more offspring from being born. This of course led to questions about the integrity of the animal Herman versus the other interests and to questions about other types of control to prevent new

---

[#] Animal Production Systems Group, Department Animal Sciences, Wageningen University, P.O.Box 338, 6700 AH Wageningen, The Netherlands. E-mail: Akke.vanderZijpp@wur.nl

offspring. The case also shows that application of the utility concept in practice is complicated.

This framework for evaluation has been adapted over time; it shows an ethical concept of human–animal relationships in biotechnological research that goes beyond the minimalist concept of human autonomy and self-determination. A holistic concept for valuation was designed in which also the position of animals versus human interests is valued, as are the interests of the environment and nature.

Next, Düwell combines the idea of the contractualist perspective of the world with human rights and autonomy. Together this will most likely lead to a minimal consensus in a world with plural moral convictions. This minimalist approach leads to questions about social ethics; what are the legitimization and the content of the moral rights of the individual, based on autonomy and self-determination? Regarding execution of research the methods can easily be discussed and the rights of the proband can easily be respected. More difficult is the insecure and unknown outcome of the research. The liberal position means that full access to information has to be fulfilled and the final outcome of research can be refused. The advantage of this approach is that it can reconcile different worldviews and works for different actors. The above case of Herman illustrates how the perspective changed over time depending on the outcome of research results and finally led to the decision to castrate the transgenic bull Herman.

But there appears to be some uneasiness with this liberal position, both in terms of the room for making decisions predetermined by the scientific community and in terms of ethical attitudes like virtue, perfectionism and care. Somehow the rights and obligations humans have towards others (humans and animals and nature for example) come back into Düwell's perspective. And the right to be part of the process to organize potentially different outcomes of research in a participatory process with scientists is essential.

The above framework of evaluation of animal biotech research appears to be much closer to the rights of autonomy and self-determination in a social perspective, viz., where the relationship between humans and animals has been defined in the context of nature and environment and the role of humans in the democratic process regarding research design.

Another case: Workshop on Foot and mouth disease in 2002

The outbreak of FMD in 2001 in The Netherlands caused country-wide protests against killing healthy, but usually vaccinated animals. The outbreak involved about 1000 commercial farmers and 1800 hobby farmers. It involved values like utilitarism, stewardship and integrity/intrinsic value of animals. The Workshop was organized beginning with a stakeholders meeting, resulting in research priorities, followed by an interdisciplinary meeting of scientists. Gaps appeared in knowledge of social science, public administration, communication and crisis management. The result has been an agenda for future research and some policy suggestions.

Characteristics of the meetings: High consciousness of own autonomy, lack of basic knowledge of veterinary epidemiology, value pluralism both regarding animals and human interactions, high (emotional) interest in sharing experiences and capability to discuss a joint approach to research, need for quality facilitation. This Workshop also showed that the concepts of autonomy and self-determination are highly developed, yet the need for sharing experiences and participatory development of future scenarios was equally important and highly valued by the participants.

Conclusion

My conclusion: To resolve complex problems in our society the liberal approach based on autonomy and self-determination rights is not sufficient. We have to define our position regarding our fellow human beings and animals, nature and the environment. See also the outcomes of the *'Waardenvolle Landbouw'* Workshop (Values in Agriculture, inVan Eck and Oosting 2001) and our research programme for sustainable animal-production systems. They appeal to a holistic approach in science, which requires social interaction for decisions about trade-offs between unequal issues regarding planet, profit and people.

References

Van Eck, W. and Oosting, I. (eds.), 2001. *Naar een waardeNvolle landbouw.* Taskforce Waardevolle Landbouw, Wageningen UR, Wageningen. [http://www.wau.nl/pers/01/taskforce-rap01.doc]

SCIENTISTS IN SOCIETY

10a

New public responsibilities for life scientists

Michiel Korthals[#]

Introduction: A say for our mouth and new tasks and responsibilities for life scientists

"What your genes want you to eat". This was the title of an article in the New York Times of 4 May 2003 on the ways in which the life sciences will influence food and drug choices of consumers and patients in the next decade. The author, a journalist by the name of Grierson, states that our diets and prescriptions will in the future be customized; to achieve this, consumers and patients will need continuous feedback between screening agents and food and drug consultants (such as general practitioners and dieticians) for continuous update of their gene passports or health cards and for relevant advice in response to new food products, drugs or scientific developments. The message of this journalist does not differ from those of other, less popular writers on the subject: if consumers are indeed health-driven and want to postpone death, then they must allow their genes to dominate their daily lives. That means allowing interaction between genes and lifestyles, even allowing life scientists and technologists to play a dominant role in their lives. The term 'gene' is indeed a metonymical expression of the whole life-science system and industry. An issue that the journalist does not address is whether consumers in future will have a say in what they put into their mouths and the related responsibility of life scientists.

To tackle these questions, I will first outline the main developments in the life sciences during the last decade, and then discuss some aspects of the traditional concept of responsibility, which stresses the causal connections between agent and outcome. I will argue that, from a pragmatic point of view, the concept of different practices can help in delineating new grey zones between conducting research, rendering advice, screening consumers and patients, consulting the public, and prescribing and selling food stuffs and drugs. Moreover, I will make it clear that professional scientists have a public responsibility; they must build new Chinese Walls to raise the level of trust between themselves and the general public.

New developments in the life sciences: genomics

Although it took some time, the discovery of DNA by Watson and Crick in 1953 has significantly changed the disciplines of biology, medicine, chemistry, food science and agricultural science. Genomics is the broad label that covers the integration of these fields into the new discipline of the life sciences. Genomics describes the integrated application of biochemistry, microbiology and process technology for the purpose of turning the potential of micro-organisms and cell and

---

[#] Applied Philosophy Group, Department of Social Sciences, Wageningen University, Hollandseweg 1, 6706 KN Wageningen, The Netherlands. E-mail: Michiel.Korthals@wur.nl

tissue cultures to technical use. Two key components of modern biotechnology are genomics in the narrow sense (the molecular characterization of organisms) and bioinformatics (the assembly of data from genomic analysis into accessible forms). Because of its enormous potential, genomics (along with nutrigenomics) is regarded as one of the key sciences and technologies for the coming decades to improve food availability (mostly attributed to developing nations) and food quality and safety (attributed to the industrialized world). It can deliver both products and methods, for example for the analysis of food safety by delivering fingerprints of genetic activity in products and in humans.

With genomics and nutrigenomics, the sharp distinction between food and medicine falls apart, and a grey zone emerges between the two. An understanding of plant-biochemical conversion processes, along with knowledge of how humans metabolize foods, will bring prevention to the centre of medicine and food sciences, shifting the emphasis from health care to healthy living. Food acquires the characteristics of medicines and determines what kind of medication is necessary; medicines become food or influence food intake. The new grey zone where health care and nutrition meet is the battlefield where social, economic, political, juridical, educational and ethical problems are emerging, a battlefield that requires constructive social and scientific thinking, intensive public debate and corresponding technologies to come up with solutions.

As with every new science and technology, many people see benefits such as cost reduction for producers and healthier products for consumers. However, aside from cost reduction or higher prices, the new technologies may also involve both tangible and intangible costs. These costs may, for example, consist of more expensive materials, a higher level of skills needed to manage the product, or a higher risk of product failure. Genomics calls for co-operation between scientists and the general public, with successful co-operation depending to a large degree on the way citizens/consumers are able to cope with these new trade-offs in institutions (some of which are to be newly established). For food researchers, policymakers, the food industry and retailers in the genomic and nutrigenomic sector there are many uncertain factors: the knowledge claims are uncertain and different types of risk are involved. This requires risk analysis and precautionary measures. The perceptions of consumers are unclear, and the economic prospects (costs and benefits) are uncertain.

For researchers in the field of genomics and the life sciences, the customary distinctions between basic applied science and technology do not exist anymore (if they ever existed). One only has to glance through journals such as *Trends in Food Science and Technology* or *Theory in Biosciences* to see that many fundamental articles directly concern the preservation of food, the relationship between genes and obesity, or nutrient cycling and sustainability; in other words, fundamental social issues are discussed. Before the DNA revolution took place, food scientists did research into the extent to which certain ingredients were poisonous, or into the preservation of food. Nowadays, however, the life sciences are expanding into the food choices of people; they have large impact on a person's daily life, which will be increasingly organized along the networks of a gene passport or health card.

For professionals the boundaries between industry, university, and government policy are blurred. They switch easily from one sector to another. Regulations and activities that used to exist in one sector are now taken over by other sectors, like the patent system or advisory activities. For patients and consumers this makes it very unclear who is speaking, for example when consulting an expert: is he or she employed by industry or somehow connected with a government agency that has something to gain or lose? Famous cases involve the role of scientists in the anti-smoking debate or, more recently, the role of food and health scientists in the sugar debate raised by the World Health Organization (The Guardian, 21 April 2003). Other issues that are covered in the quality press are the relationship between screening agencies and the therapeutic products that are prescribed and sold (Sciona) and the management of bio-databanks (Bulger, Heitman and Reiser 2002).

It is hardly necessary to refer to famous disasters in science and technology, such as the nuclear energy plant in Chernobyl, the Challenger and Colombia space shuttles, the marketing of GM maize by Monsanto, or the extensive use of X-rays in medicine during the 1930s, to become pro-active in the regulation of responsibility and accountability with respect to new developments. Although, historically speaking, the first reflections on responsibility refer to disastrous consequences and their punishment (John Stuart Mill, Kennett 2001), nowadays we have a more mundane conception of responsibility, which stresses the potential implications of actions and technologies and not only the negative effects (Resnik 1998). Who is responsible for the widespread implications of life-science technologies for individuals, groups and social players vis-à-vis these new developments? To establish all the links in the life-sciences network, it is necessary to assess these different factors and to establish trust between the various stakeholders and with consumers in particular. However, the various stakeholders do not put the same emphasis on these issues, their approaches in solving the issues differ widely, and so the dance of evading responsibilities continues. Companies state that they only respond to what the market and consumers want them to do; life scientists argue that they only want to conduct research and do not determine profit strategies; politicians say that they are advised by scientists and have privatized these developments at the request of citizens. I consider it better to step outside this dance, and to look beyond the parties to three impartial conceptions of responsibility and accountability: the causal theory, the role theory and the pragmatic theories of responsibility.

Causal theory of responsibility

According to the causal theory of responsibility, responsibility is associated with the causal relationship between an individual person and his activities. Only when a person is in full control of the circumstances, of his agency and of the activities that he sets in motion (thus when the acting person is free), can he be held responsible. Surely, it would be unfair to make people responsible for events when they do not have control over the circumstances. According to this view, if a scientific invention is stolen from a laboratory and misused, the inventor cannot be held responsible. As a matter of fact, however, many of the acts that we perform, we do not control. There are many circumstances as well that we do not control, that we at least do not decide about and that determine our decisions.

There are several approaches to this traditional philosophical issue of freedom and determinism. On the one hand, there is the determinist approach, which denies all agency (Pereboom 2001); consequently, talking about responsibilities is superfluous. On the other hand, there is the voluntaristic view (as in the existentialism of Sartre); and thirdly, there are several approaches that stress the compatibility between freedom and determinism (compatibilism). Kant's compatibilist point of view emphasizes the agent point of view next to the deterministic view. He maintains that freedom and determinism are ways of reconstructing sequences of events. From the deterministic point of view, we organize reality according to relationships of causes (physical and otherwise) and effects, with humans merely being effects (i.e. according to the laws of nature). From the perspective of freedom, we view moral agents acting as free causes according to moral relationships between intentions and acts (i.e. according to the laws of freedom). What this perspective does not take into account, however, is that intentions are shaped by cultural and social circumstances as well, that they are neither totally in control nor totally out of control of the agent. Actions can be caused by something beyond our control, while we can still be in control of them. This makes the perspective of freedom and responsibility more complicated. In fact, it urges us to reconstruct the distinction between freedom and determinism not as one between two different and excluding viewpoints, but as a gradual distinction: an agent can be more or less free, more or less determined.

Role theory of responsibility

Linked to this complexity is the issue that the description of acts generated by human agents can vary. Suppose I intentionally open the outside door of my house, simultaneously detect a prowler waiting in the dark, hurt the kid that was just trying to open the door from outside, and alarm my partner who is sleeping upstairs (what Feinberg calls the accordion effect). The social contexts that shape actions and their interpretations are not taken into account by the causal theory of responsibility. In daily life we normally apply a more social theory of responsibility, with causality being less important. When we hold parents responsible for the actions of their children, drunken drivers for their risky behaviour and politicians for their civil servants, this has to do with our expectations of the parental role (care) and our assumptions about driving (safety) or political roles (stewardship) and not with causal responsibility or lack thereof. According to the role theory of responsibility, we allocate responsibility to individual persons depending on their social roles and the social contexts, and not only depending on the extent to which they actually cause certain events or have a conscious say about the occurrence of these events. This allocation has everything to do with the normative structure of our societies and is not a merely empirical fact. However, the social contexts do not supersede causal accounts but complement them. Intentionally, negligently or recklessly causing harm by conducting certain types of research (remember the Tuskegee Syphilis Experiment) is still covered by this concept of responsibility.

There are several issues that make role responsibility questionable. First, take the comment of the Challenger director to the engineer who warned about the poor sealings in the rocket that ultimately caused the catastrophe: "Take off your engineering hat and put on your management hat". It is often not clear what one's role is as a life scientist or engineer in an organization or society, and I would even argue that certain role aspects (like safety) should override other aspects (like profit or management). A second issue is that, even in situations where the practice of science and technology is well organized, there is still the issue of the goals of the organization, its research agenda, its research priorities and its research design. Scientists have a responsibility to society at large as well, one that transcends role responsibility and harbours on public responsibility. We only need to recall the appeals of Einstein and Oppenheimer to ban atomic-bomb testing, or the actions of Rachel Carson and Barry Commoner in the 1960s about the banning of DDT and in favour of biodiversity, to recognize that scientists have a certain responsibility towards the common good, a responsibility that transcends their specific role when working in a certain organization and doing their job (Shrader-Frechette 1994). A third issue is that roles are subject to change, along with the practices that they are part of. With respect to life scientists, when their practices turn to social life as they become involved in gene passports or health cards, their social role changes accordingly. Here again there is no causal responsibility, but at least a shared responsibility for large-scale effects. But to what extent and in what measure?

Pragmatic theory of responsibility

In several publications we have presented a revised version of pragmatic ethical theory, in which we first stress the three antis of pragmatism: against foundations of moral guidelines in metaphysical or other entities; against common dualisms like nature and culture, citizen and consumer, mind and body, science and ethics; and against fundamental doubts à la Descartes (doubting the validity of everything) or total societal critique (doubting the social system).

The first 'anti' means that I am sceptical of general principles such as the four justified by Beauchamp and Childress (the principles of autonomy, justice, maleficence, and beneficence). Only as heuristic guidelines can they play a role in ethical reasoning. Secondly, we stress the importance of values in a globalizing world, such as democracy and co-operation, in other words, peaceful ways of managing ethical and other normative differences. Finally, we delineate the concept of co-evolution of technology and ethics, in the sense that new technological developments are seen as intriguing challenges for common morality frameworks; they do not function a priori as impulses to draw boundary lines or erect red stop signs. Co-evolution of technology and ethics makes it clear that both change when reacting upon each other. In that sense, technology has a broad ethical component, and ethics is intrinsically connected with technology; neither can be held constant and unchanging (Keulartz et al. 2002). However, we should be wary of ethical colonialism, in the sense that ethical problems are considered to be rampant, with the consequence that scientists are overburdened with all the moral problems of the world.

In analysing urgent ethical problems (problems that hurt, not those that may arise but are rather far away or pure science fiction), I use the concept of practices, in particular their interrelationship and their relationships with public debates, consultations and decision making. This concept is useful because it clarifies how technologies are themselves part of social practices that are applied to other social practices. Embedding life sciences and their corresponding technologies in social practices means therefore to open up the social practices of science to the social practices to which they are applied and looking for positive connections between these practices and negative, controversial encounters. For example, when genetics started to use new technologies to predict and treat certain diseases such as Huntington, patient organizations, hospitals, clinics and advisory agencies changed correspondingly, as well their standards of excellence and broader norms.

In searching for compromises and new possibilities (scenarios), we pragmatics try to re-open the frozen frontiers between practices. Practices have values and goals, like standards of excellence, as well as concrete products that are measured against these values and goals. When practices change, the norms change as well, which is not only of interest for the practices involved but for society as a whole. That calls for public consultation.

Recent developments in the life sciences put them into new constellations with other practices of health care and food, in that they change these practices and require that many traditional standards of excellence and broader norms should be made subject to revision. Because of the huge investments involved, private–public co-operation ('triple helix', Etzkowitz and Leydesdorff 2002) is growing more common. Regulations that are maintained in the private sector, such as patents, are becoming common in the public sector, including the universities. Other regulations and attitudes are moving from public sectors to semi-public sectors. Health diagnostics, health screening and health consulting, which used to be performed only in the health sector, now are also regularly performed in the food sector, even though the necessary moral and legal regulations are often lacking.

Two issues are in my view of utmost importance to be discussed by professionals, both in university and in industry settings. First, there is the topic of research priorities. If universities and their laboratories are indeed so heavily dependent on financial support by industry, who then chooses the research topics and why? Are the illnesses for which cures are sought indeed illnesses that should be urgently treated because many people suffer from them, or are they the luxury complaint of a rich minority? In the ivory-tower university environment of the early twentieth century, setting the research agenda was the privilege of peers, but nowadays this is no longer the case. Do governments decide, is it industry or the scientific community, or is it some unclear combination of the three (Nestle 2000)? Should these decisions be preceded by public debates and followed up afterwards by public scrutiny so that decision makers can be held accountable? What research topics are societal and scientifically relevant, and in what way are they linked to social goals that are thought to be relevant according to the general public? Who decides which alternatives should be chosen from the possibilities, given the present state of the art of the life sciences? What standards should be applied (Kitcher 2001)? Nowadays, these more general moral requirements of responsible scientists, like working for the public good (Shrader-Frechette 1994) are not clear, simply because it is not clear what the common good is in, say, the case of health food or food safety (how much unsafe food is acceptable?). Here only public debate and consultation can help both professionals and the public at large to find orientation. The main public responsibility of life scientists in deciding on a certain research project is therefore not that of giving information on fraud or on discrimination (although both actions are as necessary as ever), but of participating in public debates, giving both information and normative guesses about the possible benefits and detriments of new developments. This activity does not require professional codes but the skills to perform open, honest and rational debate. However, this can lead to a conflict with the second issue that ought to concern life scientists.

The second implication for life scientists in their new social constellations is that far more than in the past, they are professionally involved in consulting private persons or private companies, in advising governments and health organizations, in managing data acquired from screening tests and diagnostics, even in proposing new foods or drugs that appear to correspond with the diagnostics. The danger is even of confusing and blurring the dividing lines between these activities. It is so easy, when you get your marvellous, expensive diagnostic tool from a certain company, also to prescribe the drugs or foods that are delivered by this same company. But is it in the interest of the consumer and the science system at large? The interests of organizations never fully coincide with the common good, whatever that may be. Will the public trust a science system that naively or intentionally assumes that its particular interests coincide spontaneously with the common good? I do not think so, and the general opinion in most western societies does not assume that the interests of a subsector coincide with societal interests. Viewed in this way, life-science professionals should set up Berlin Walls between conducting research, giving advice, doing diagnostics, public consulting, and prescribing foods or drugs. In giving advice to a consumer or patient, the professional must clearly state in advance what his potential conflicts of interests are. I am ethically pleased to see that this is already a common practice in articles in some high-ranking medical journals. In food science, however, it is still very uncommon in publications, let alone in contacts with consumers or policymakers. But this could be done much more radically, in the sense that scientists who conduct research that is financed by industry indeed refrain from giving advice to consumers and from prescribing certain foods or drugs, that they do not participate in public debates, and that all these activities are neatly, physically and personally, separated in the same way that banks and financial investment companies now separate their different activities. Such separation may be referred to as a Berlin Wall.

However, from the point of view of the urgent need for life scientists to participate in public debate and policy structures, Berlin Walls are too cumbersome, as they can prevent the necessary flow of information and communication between the various activities. So maybe Chinese Walls between conducting research, public consulting, advising and prescribing are a better solution from a pragmatic view. It is not forbidden to change sectors, but only on certain conditions, like announcing your conflicts of interest, outlining possible losses when you go along with this particular scientist, and honestly indicating alternatives. For example, in the sugar debate it should be a matter of professional honesty that, when a professional speaks out for the rules of the World Health Organization or against them, he makes his connections with industry and government clear in advance. Also, if a scientist is against regulation of sugar, he should state what alternatives there are to curb sugar intake and reduce obesity. Again, professional codes do not help here, because they mostly deal with the avoidance of harm and with honesty, and not with learning to find out where these conflicts of interest arise with the norms of other people such as various consumer groups and with other ethical skills and competencies. These kinds of skills add to the public responsibility of life scientists, which involves participation in a rational and decent way in public debates and in transparent decision making.

Conclusion

In this paper I have discussed several types of responsibility for life scientists in the new constellations, where the boundaries between sciences, advising, managing information, prescribing and profits are becoming blurred. In the light of shifting commercial and political conditions and of changing relationships between medical and food practices, I have outlined new tasks and ethical issues. From the pragmatic point of view I have presented two central issues that are to be dealt with. The first issue is that of setting research priorities and their relation to the public at large. Considering that the public has an eminent interest in the new life-science developments but simultaneously does not understand them, the public responsibility of life scientists should be cultivated by training them in making rational conjectures about future possibilities of the combination of scientific development and social change. This is difficult, because scientists should refrain from publishing immature or non-validated results.

The second issue is even more complicated. Yes, we need certain Chinese Walls, not Berlin Walls, between the roles of research for universities, governments or industries, of public consultation and of advising industry or patients/consumers. The new tasks of the life scientists require more than ever that there is total clarity in whose interest they conduct research, construct and manage data banks, prescribe foods, and so on. According to a pragmatic view of ethics, the trust of consumers in the newly establishing life-science network depends upon the question: will consumers in future have a say in what they consume? The public responsibility of life scientists should be reconstructed with that question in mind.

References

Bulger, R.E., Heitman, E. and Reiser, S.J. (eds.), 2002. *The ethical dimensions of the biological sciences*. 2nd edn. Cambridge University Press.

Etzkowitz, H. and Leydesdorff, L. (eds.), 2002. *Universities and the global knowledge economy: a triple helix of university-industry-government relations*. Continuum, London. Science, Technology and the International Political Economy Series.

Feinberg, J., 1970. Action and responsibility. *In:* Feinberg, J. ed. *Doing and deserving: essays in the theory of responsibility*. Princeton University Press, Princeton, 119-151.

Kennett, J., 2001. *Agency and responsibility: a common-sense moral psychology*. Oxford University Press, Oxford.

Keulartz, J., Korthals, M., Schermer, M., et al. (eds.), 2002. *Pragmatist ethics for a technological culture*. Kluwer Academic Publishers, Dordrecht. The International Library of Environmental, Agricultural and Food Ethics no. 3.

Kitcher, P., 2001. *Science, truth and democracy*. Oxford University Press, Oxford.

Nestle, M., 2000. *Food politics: how food industry influences nutrition and health*. University of California Press, Berkeley. California Studies in Food and Culture no. 3.

Pereboom, D., 2001. *Living without free will*. Cambridge University Press, Cambridge.

Resnik, D.B., 1998. *The ethics of science: an introduction*. Routledge, London. Philosophical Issues in Science.

Shrader-Frechette, K.S., 1994. *Ethics of scientific research*. Rowman & Littlefield, Lanham. Issues in Academic Ethics.

Watson, J.D. and Crick, F.H.C., 1953. Molecular structure of nucleic acids: a structure for deoxyribose nucleic acid. *Nature,* 171, 737-738.

10b

Comments on Korthals: New public responsibilities for life scientists

Jan H. Koeman[#]

First of all I would like to comment on some of the factual considerations with regard to the new revolutionary developments in the life sciences, viz., the booming field of genomics, which forms the basis for this paper. Secondly, some remarks will be made on "Role theory of responsibility" and finally there will be some comments made on the main message of the paper, namely: "New public responsibilities for life scientists".

Factual considerations

Genomics and related fields such as proteomics and nutrigenomics, are booming areas indeed. However, to state that because of the enormous potentials, genomics (and nutrigenomics) is one of the key sciences and technologies for the coming decades to improve food security and food quality and safety, may finally turn out to be an exaggeration.

It should be noted that until the 19th century almost nothing was known about the composition of all the different kinds of food and the nutritional requirements of the human being. The present insight into the nutrient composition of food and the nutritional requirements of man are the results of impressive scientific developments in fields like human physiology, food chemistry, biochemistry and cell biology, which took place in the last century. Think about our present in-depth knowledge on the physiological functions of essential nutrients, such as carbohydrates, proteins, lipids, amino acids, vitamins and minerals. Important deficiency diseases became understood and appropriate guidelines could be provided to consumer populations in order to prevent such diseases. In the meantime it has been shown that the onset of many other, mainly chronic, diseases, such as forms of cancer and atherosclerosis is also influenced substantially by diet and nutrition. Also in these cases the adjustment of nutritional behaviour and the modification of the dietary composition have beneficial effects on the overall incidence of such diseases.

The suggestion now is that the genomic age will lead to revolutionary improvements in the prevention of nutrition-related disorders, in addition to the achievements that have already been made. For instance, that each consumer will have his or her gene passport "enabling the unique tailoring of future diets" as stated recently by a journalist of the New York Times (quoted by Korthals in this volume). However, the genetic background of health and disease later in life is very complex in the sense that it is expressed through the interaction of a complex of genes. This makes the predictive value of 'genomics' very low on an individual basis. There are only few cases where a disease or disorder is related to one or a few dominant genes

---

[#] Toxicology Group, Wageningen University, Thijsselaan 5, 6705 AK, Wageningen, The Netherlands

or chromosomal aberrations, for instance Down's syndrome or haemophilia. Certainly 'genomics' will lead to important improvements in the ability to identify individuals with deficient or anomalous genetic properties, but these will be rather exceptional. Therefore it is most likely that advice to future consumers will mostly be addressed collectively rather than individually.

In my opinion the statement in the paper that "the sharp distinction between food and medicine will fall apart" is also questionable. The era of genomics is just another step ahead in food and nutritional sciences.

Role theory of responsibility

However, my comments are of no concern for the remarks made by Korthals on the position of researchers in this field. For professionals the boundaries between industry, university and policy were already blurred in the pregenomic age. For decades they have already jumped from one sector to another, and the quality of the communication between scientists and consumers is far from perfect. Therefore Korthals's plea for a reconstruction of the public responsibility of scientists stands.

I think it is interesting to note Korthals's views on role responsibility. He states that there are at least two issues that may make role responsibility questionable.

The first is that the role of a life scientist or engineer is in an organization or society is not always clear. He argues that concerns of certain roles (like safety) should override other aspects (like profit and management). I can give you an example from my own discipline, toxicology, to illustrate this point. Normally in pharmaceutical and chemical companies the toxicologist in chief directly reports to the board of directors. If the risk assessment turns out to be negative within the context of the anticipated use of the product, the board will generally accept the negative advice given by the toxicologist. A colleague of mine was in such a position in a multinational company. He had a respectable position until there was a major change in the management structure. Shortly thereafter he refused to approve a newly developed product because he felt its use would pose a human-health risk. He was overruled and had to retire from his position. Fortunately such events are very unusual. Apparently at that time role responsibility was not taken seriously by the management of this company.

The second issue which may make role responsibility questionable relates to the position a scientist has with regard to the goals of the organization, its research agenda, its research priorities and its research design. Here Korthals correctly states in my opinion, that scientists have some responsibility to society as well. Also in this case I can refer to some personal experience. In 1966 we were the first ones in Utrecht to report the presence of PCBs (polychlorinated biphenyls) in wildlife tissues. It appeared that we were just second in the world. A Swedish scientist was the first to publish the discovery of these compounds in the New Scientist. We wrote to three chemical producers of PCBs asking their scientists for detailed scientific background information. Number one, a German company, did not reply. The second, a French company, replied that what we had found must have been artefacts, because it was supposed to be most unlikely that PCBs would turn up in wildlife samples. However, the third, an American company, responded adequately. Two of their responsible scientists came to our department to discuss our findings and were immediately convinced. A few years later this American company was the first company to stop voluntarily with the production of PCBs. Apparently only in the American company scientists had the mandate to discuss the scientific merits of the problem freely with people outside the company, thus overriding the immediate commercial interests.

New public responsibilities for life scientists

Two issues are mentioned under this heading in Korthals's paper. The first one refers to research priorities, the second to what I would like to call the containment of the different scientific duties a scientist may have; The Berlin Wall between doing research, giving advice etc. as described by Korthals.

He states that within the framework of setting priorities life scientists should participate in public debates by giving both information and normative guesses about the possible benefits and losses of the new development. I would prefer 'costs' instead of 'losses'. I feel this already happens in many respects both generally in panels and through the media. But it could be improved. For instance, scientists could improve their abilities to communicate with the media. In the US the Foundation for American Communications (FACS) has published a media guide for academics which is very helpful (Rodgers and Adams 1994). But improvements could also be made within regulatory frameworks. A few years ago a panel of governmental scientists from the UK, Sweden, Denmark and The Netherlands proposed a framework for an integration of risk analysis and trust that comprises the involvement of expert panels as well as citizen's panels as shown in Figure 1. The scheme has not yet been approved officially but the principles are increasingly applied informally in various countries. This model deserves to be given a wider application to stimulate debates between scientists and the public in other fields of science as well.

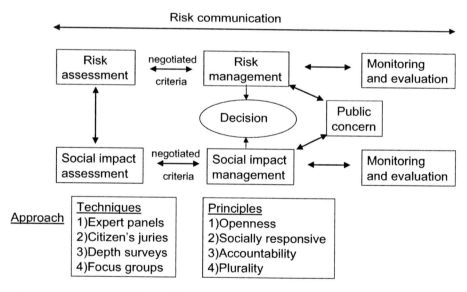

Figure 1. Integrated risk analysis and trust (slightly modified version of the scheme presented by Barling et al. 1999)

However, irrespective of the model chosen the public responsibility of the scientists will mainly remain a matter of personal integrity. If they cannot meet the appropriate standards in this respect they should not be allowed to cross Korthals's Chinese Wall.

References

Barling, D., De Vriend, H., Cornelese, J.A., et al., 1999. The social aspects of food biotechnology: a European view. *Environmental Toxicology and Pharmacology*, 7 (2), 85-93.

Rodgers, J.E. and Adams, W.C., 1994. *Media guide for academics*. Foundation for American Communications, Los Angeles.

11

Science, context and professional ethics

Ruth Chadwick[#]

"There is a central core of universal values that any truly modern society must possess, and that science promotes. These are rationality, creativity, the search for truth, adherence to codes of behavior, and a certain constructive subversiveness" (Serageldin 2002)

Science and the ethics of distrust

In the last decade ethical issues relevant to scientists as professionals have come very much to the fore, although they have not, typically, been considered under the guise of *professional* ethics as such. Rather, appeal has been made to problems that have arisen in the context of, for example, genetically modified foods and the BSE case. These cases involved both public concerns over the profit motive prevailing in the research and policy agenda; and anxieties about the unpredictability of long-term consequences. Can the framework of professional ethics shed any light on these issues?

It might be relevant to consider the wider context of distrust in the professions (Pellegrino 1991). Both sociological and philosophical criticism (cf. Koehn 1994) have constituted aspects of this phenomenon, apparent in the late 1980s and early 1990s. An ethic of distrust proceeds by attempting to regulate more closely the activities of professionals, by increased external monitoring and demands for accountability. According to one view (Veatch 1991) this approach wins the day by default because the notion of an ethics of trust is not only difficult to sustain: it is actually incoherent.

Robert Veatch attacks what he sees as the three arguments supporting an ethic of trust: (1) that professionals serve the client's interest; (2) that professionals can present value-free facts to the client; (3) that professionals should act on a set of virtues inherent in the profession. Veatch argues that modern professionals ought not to know what the client's interests really are – the most they can know is what the client's interests are in one particular area of life. Whereas medical professionals might be primarily concerned with promoting health, for example, health might not be the top priority for a patient (Goldman 1992). Veatch further argues that professionals cannot present value-free facts; and that it is a serious mistake to think that any given profession is associated with one particular conception of virtue. These points are very apposite for a consideration of the ways in which ethical considerations enter debates about science, in the light of, for example: the questioning about the extent to which science serves society; debates about value neutrality in science; and the extent to which scientific 'truth' is an unquestionable good. But does science constitute a 'profession' in the relevant sense to make it worthwhile looking at it through the lens of professional ethics?

[#] Lancaster University, Furness College, Lancaster LA1 4YG, United Kingdom. E-mail: R.Chadwick@lancaster.ac.uk

Chapter 11

Science as a profession?

There are different approaches to giving an account of what it means to be a profession: the 'defining characteristics' approach and the 'process' approach. The 'defining characteristics' approach, again, has a narrow and a wide version – in a wide sense 'profession' simply means someone's occupation; in a narrower (more interesting) sense it refers to a certain kind of activity, one carrying with it a certain status and associated with a particular ethic. Traditionally a profession has been marked out by a body of knowledge, mastery of which (at least partly) regulated entrance to its ranks; and by an ideal of service (Airaksinen 1994). Since the body of knowledge has the potential to confer power, money and status, professionals are expected to use their skills for the benefit of the community. Those groups which have long been secure in their recognition as professions, the so-called liberal or 'learned' professions such as medicine, divinity and the law, have also been characterized by a considerable degree of authority and autonomy in their practice. Along with the autonomy of the individual professional, professional bodies have also been accorded a significant degree of autonomy in controlling both accesses to the profession and professional conduct.

Taking into account the provisional nature of scientific knowledge, science as a profession can apparently satisfy the 'body of knowledge' criterion: the ideal of service, however, is less clear. Service to whom? This point may depend on the context in which the scientist works, and this is a very important consideration with regard to a discussion of the ethical issues. One of the problems, arguably, with current scientific practice is the rival claim of academia, industry and government as the context in which scientific research takes place.

If we look at a process account of profession, in terms of how an occupational group achieved a certain status (cf. Freidson 1994), rather than the set of characteristics approach, it is arguable that scientists have achieved a position of power, not only in having far-reaching effects on society through scientific advance itself but also in having considerable influence as government advisers and being able to command the highest salaries among academics.

Traditional classifications of professions have been subject to two contrasting trends: first, the attempt by some groups for recognition as professions or neo-professions; and second, challenges to the notion of professionalism either because of its conceptual inadequacy, or on the grounds of its social consequences. I shall set aside the former for present purposes, but I do want to say something about the second. One reason for the challenge to the concepts of profession and professionalism is that critical, reflective professionals, with autonomy over their practice, may be seen as a threat (Williams 1996). A second reason is connected with the potential for professions to become self-serving elite's (Freidson 1994; Illich 1997). In response to this situation there has been an attempt to replace the focus on professionalism with a focus on competencies. It may not be so easy, however, to dispense with the notions of 'profession' and 'professionalism'. J.K. Davis, for example, has argued that for the professional it would not be sufficient that a client was satisfied, if the professional him- or herself felt that the service was below standard. For the professional, however, it is more than simply doing a competent job: a worker becomes a professional by professing reasons for doing their work in a certain way (Davis 1991a).

We cannot assume, however, that the area of 'competence' is ethically neutral, while values come into the realm of 'reasons for action'. Certainly in the case of science, while there may be dispute about what we mean by calling science a profession, it is increasingly recognized that while *scientific* competence may be necessary it is not sufficient for the 'good' scientist. A recent article in *The Daily Herald* (2003??) said:
"Here are three biology terms: endoskeleton, enzyme, epidermis. If you're serious about a career in biology, add one more item to the list: ethics". Here ethics appears among the basic *competencies* for a biological career.

What I want to suggest is that it may be enlightening to consider science as a profession and to look at it through the lens of professional ethics. It enables us to put aside the specific 'scandals' that have, supposedly, given rise to the distrust of science and to look at the wider context of the distrust of professional power, the reasons for it, and the proposed solutions, such as the approach to grounding trust in the professions anew. Koehn (1994) has argued that to do this is important because professions represent the mechanism chosen by Anglo-American morality for providing people with goods such as health and justice, and if professionals are not trustworthy, where are we to turn for help? For health and justice are not goods that are readily dispensed with. Does the same apply in the case of science? This depends on identifying the relevant good. For Koehn, the challenge is to show not only that there are grounds for trust in the professions because they provide people with such goods, when they lack them, but also that they do not violate the requirements of ordinary morality (Koehn 1994). There is a connection here with current policies of trying to re-establish trust in science by, for example, the developments we have seen starting from public understanding of science, moving through public consultation, to public 'engagement'. At what stage in the process should the public be engaged? There has been debate about the limitations of involving the public only at the 'downstream' stage of the impact, implications or applications of science, rather than at the 'upstream' stage of debating what scientific research should be carried out. It would be possible to move even further back, however, to consider what the 'good' of science is: is that for scientists themselves to determine?

Although it seems that there are close connections between the debates in professional ethics as a whole and those concerned with the ethics of science, it is clear that *context*, as we have already noted, is important. In so far as scientists are academics, the relevant questions of professional ethics will be common to other academic disciplines – I am thinking here of the avoidance of plagiarism, for example. In this paper however I want to consider whether and to what extent there are issues that are specific to science and the relevance of context in addressing them. With that in mind I shall move on to the problems of professional ethics.

Professional ethics and science

Problems of professional ethics typically fall into two broad categories, but both arise essentially from professional power. The first is concerned with the professional–client relationship, while the second relates to the role of professions and professionals in society as a whole.

The professional–client relationship

Although an ideal of service is supposed to provide a safeguard to promote the use of professional expertise to help rather than harm, specialist knowledge, to which professionals have access and clients do not, does give the professional power, and the client is thus placed in a vulnerable position. One caveat however (Langan 1991) is that the paradigm of a relationship between two individuals is inadequate because it overlooks those professions which do not, or not substantially, conform to this pattern, such as teaching, which may be but commonly is not done on a one-to-one basis. Although there are ethical issues arising in relation to science concerning treatment by researchers of individuals, e.g. in human subjects research, science *per se* also does not conform to this pattern, having as it does an impact on society as a whole.

The second category of problems is more concerned with the role and image of professionals in society. While it may be true (Pellegrino 1991) that there has always been a tendency towards distrust of professionals, this has been exacerbated by social and political developments. The trend towards client autonomy; attempts by government to curb the independence and privilege of professionals; media criticism have all had their effect.

Professions and science in society

Sociological critique has suggested that professions, rather than being essentially moral enterprises, are in fact effective monopolistic institutions and that the professed commitment to ethical ideals, rather than conferring legitimacy on the profession, is nothing more than ideology. Ivan Illich (1997) famously termed the mid-twentieth century the age of 'disabling' professions: far from using their knowledge to serve, they had become forms of control, claiming the authority to determine human needs. This critique is one that has been levelled against science, leading to calls for the democratization of science.

There are several aspects to this criticism of science:
(a) the belief that scientific progress is inevitable is under challenge, and indeed, that there is such a thing as progress
(b) the attempt to draw a distinction between the pursuit of knowledge and questions about its use has been undermined – it is no longer adequate for the scientist to say 'I just do the science: it is for society to decide what to do with the knowledge'
(c) perceived undesirable consequences of scientific developments 'going too far' have led people to fall back on ideas about the natural and familiar.

What I want to suggest is that while the tendency has been to address these questions by trying to make science and scientists more accountable, this has been inadequate. We have seen in the last ten years moves in many countries to do this in a number of ways, for example by allowing other forms of expertise, such as ethical and lay expertise, to influence debates in the policy area. This has had the effect of opening up the whole notion of expertise and what counts as a relevant 'body of knowledge' for particular purposes. The approach, however, has been what I call 'external' and again I think here the discussion that has taken place in professional ethics about internal and external ethical approaches might offer some useful insights.

Theoretical perspectives: internal and external

A self-derived ethic?

Some critics have taken issue with a self-derived ethic which permits professionals to be guided by standards other than those of ordinary morality. "Problems in professional ethics typically arise when the values dominant within particular professions come into conflict with other values in the course of practice.

Professionals are likely to perceive these values as dominant where others may not" (Goldman 1992, p. 1018). While few might subscribe to the view that nothing else can compete with the value of a new fact (quoted in Vyvyan 1971), a self-derived ethic might take a number of forms. In one form it is associated with the idea that there are certain ways of behaving appropriate to different roles, which diverge from those suited to people who do not fill that role. For example, it might be argued that a lawyer is under an obligation, arising out of the lawyer's role, to achieve the best result for a client even if that conflicts with what he or she believes as a private individual.

Another form which a self-derived ethic might find expression in is a code of professional conduct or code of ethics. The possession of a code of professional conduct has been pivotal in debates about what constitutes a profession. Such a code can fulfil a variety of functions: offering a public statement of ideals and values; providing a disciplinary mechanism for a professional body; reassuring the public that the profession upholds certain standards; and educating members of the profession to 'think like' others in the group (Davis 1991b).

The standards incorporated in a code may be either higher or lower than the standards of ordinary morality. Professionals have traditionally been prevented from doing things which people in other spheres of activity are permitted to do e.g., advertising. This arises out of the purported commitment to serve first the interests of clients, rather than their own profit. On the other hand this same commitment can act as a shield to protect professionals from the criticism that they do things which would be frowned on in terms of ordinary morality e.g., lying to clients or physically hurting them in order to promote some further end identifiable as being in the client's interests (Häyry and Häyry 1994).

Criticism of a self-derived professional ethic, whether in the form of role ethics or a code of conduct, is based on arguments that if an action is morally right it should be susceptible of justification by the same moral arguments that apply to the behaviour of any other member of society - professionals should not require special ethical norms to be determined by themselves. For it is not clear how such norms could be justified if not by common moral principles (Goldman 1992).

How would these considerations be applicable to science? There have been attempts to outline sets of ethical principles for scientists. One example is the HUGO Ethics Committee Statement on the Principled Conduct of Genetics Research (1996). Drawn up as it was by the Ethics Committee of the Human Genome Organisation, it is not entirely self-derived because the Ethics Committee members are not all members of HUGO, but its primary audience is scientists who are members of that organization, and who are engaged in genetic research. (It is worth mentioning however, that it is to a large extent in the context of genetics that recent debates about the ethical conduct of science have been situated). This statement is sometimes described as a 'Ten Commandments' or the 'Ten C's'. I shall not enumerate all the principles. As the statement relates to human genetic research, several of the principles relate to treatment of research participants, and I want to confine myself to science *per se*. The first principle concerns competence, which is said to be an essential prerequisite for research, and which has been mentioned above. Others which I think are relevant to the present discussion relate to *communication, collaboration* and *conflict of interest*.

Communication is relevant to being 'accountable' (cf. Holdsworth 1994) but the HUGO Ethics Committee states that "Communication is a reciprocal process; researchers must strive to understand as well as to be understood". It is stated that

"[C]ollaboration ... in the free flow, access, and exchange of information is essential not only to scientific progress but also for the present or future benefit of all participants". It is this principle of ethical science that is held to be under threat from scientists' loyalty to particular organizations, and this explains the importance of the principle stating that "any actual or potential conflict of interest be revealed at the time information is communicated and before agreement is reached".

Internal goods

A more interesting distinction between internal and external perspectives is that between internal and external goods. The internal approach might attempt to derive values internal to specific professions by examining the *point* of those professions, or the relevant *good* they produce, as outlined above. Rather than accepting them as Illich's 'dominant' professions that take it upon themselves to define human need, the question to ask is: what pre-existing human need or value do and should they serve? This quest might take different forms. The identification of health and justice as goods that cannot readily be dispensed with, because they may be needed by vulnerable people, has been mentioned (Koehn 1994). Or there might be an argument for some intrinsic or 'transcendent' values embedded in a professional activity (Tur 1994). Thirdly, knowing the point of a practice such as a professional activity might point the way to virtues internal to the practice of that activity.

Does this sort of analysis make sense in relation to science? What might qualify as the 'internal good' of science in this sense? The European Group of Advisers to the European Commission in its 1997 Opinion (Group of Advisers on the Ethical Implications of Biotechnology GAEIB 1997) referred to "the fundamental principle of freedom of research, which flows from freedom of thought". It is difficult to accept that freedom of thought can be the relevant internal good, however. Even if there are grounds for thinking that this is a good in itself, it surely cannot be the relevant good in terms of professional ethics. It is not specific to science, and it is not clear, without more, why it should be a service to the community. The relevant internal good must be something that is provided for those *affected* by the profession rather than a good *to members* of the profession.

The internal good that science provides must be in some way connected with the purported benefits to society that science can provide. If the matter is looked at in this way, it becomes clear why there are demands for the public to be engaged at a more 'upstream' stage, rather than only after the event, because arguably there are some categories of research that, for social reasons, should not be done. The European Group of Advisers (Group of Advisers on the Ethical Implications of Biotechnology GAEIB 1997) argued that the freedom of thought had to be reconciled with the protection of European citizens and human responsibilities towards animals and the environment, but this is far from being confined to the conduct of research. The very *decision to undertake* certain research might express discriminatory attitudes, for example, as in research on the genetic basis of homosexuality.

Context

My argument is that this dimension of the debates about the ethics of science has been overlooked, and that an investigation of it could help us in addressing specific problems about the context in which science is practised. The issues of context, it seems to me, are two, related to money and power. Now clearly money and power have been issues in professional ethics generally, especially in the 'process' account of professions and the sociological critique of professionalism. In science however they take on a particular character. First there is the debate about commercialization, which concerns scientists working for profit-making organizations and the pressures to which that might lead. The second concerns power, and the role of scientists on 'expert' committees.

Attempts have been made to address these issues via the external approach. For example, the profit-making issue has been addressed using concepts such as benefit-sharing, as in the HUGO Ethics Committee Statement on Benefit-Sharing (2000). This statement made the fundamental point that there are issues of justice to be addressed here, partly in so far as benefits accruing from scientific research are frequently relying on publicly funded resources to make private profits. As already mentioned, the power issue has been addressed through various mechanisms of public involvement.

However, what needs to be examined is the extent to which different institutional contexts are at variance with the 'internal good' of science, the very point or rationale of the activity. The quotation at the beginning of this paper suggests some universal values that science *promotes* and which are said to be essential to any truly modern society. To take one of these, constructive subversiveness, it is easy to see that some institutional contexts which require loyalty to the institution would be incompatible with this and which can lead to disaster (cf. Davis, M.K. 1991??).

Conclusion

I have argued that current debates on the distrust of science have missed what might be an enlightening dimension, that is to set the debate within the context of professional ethics as a whole. Reference to this context shows that it might be helpful to contrast the internal and external approach. Present day debates about the ethics of science, while trying to incorporate public engagement 'upstream' could usefully be informed by discussions about what constitutes the 'internal good' of science. This should not be understood purely in terms of freedom of thought. Analysing this good would also provide a framework for analysing the problems arising from scientific research in particular institutional contexts which might by their very nature undermine the pursuit of the internal good.

References

Airaksinen, T., 1994. Service and science in professional life. *In:* Chadwick, R.F. ed. *Ethics and the professions.* Aldershot, Avebury, 1-3.
Davis, J.K., 1991a. Professions, trades and the obligation to inform. *Journal of Applied Philosophy,* 8 (2), 167-176.
Davis, M., 1991b. Thinking like an engineer: the place of a code of ethics in the practice of a profession. *Philosophy and Public Affairs,* 20 (2), 150-167.
Freidson, E., 1994. *Professionalism reborn: theory, prophecy and policy.* Polity Press, Cambridge.
Goldman, A., 1992. Professional ethics. *In:* Becker, L.C. ed. *Encyclopedia of ethics.* St. James Press, Chicago, 1018-1020.
Group of Advisers on the Ethical Implications of Biotechnology GAEIB, 1997. *The ethical aspects of the 5th research framework programme.* Office for Official Publications of the European Communities, Luxembourg. European Group on

Ethics Opinions no. 10. [http://europa.eu.int/comm/european_group_ethics/gaieb/en/opinion10.pdf]

Häyry, H. and Häyry, M., 1994. The nature and role of professional codes in modern society. *In:* Chadwick, R.F. ed. *Ethics and the professions.* Aldershot, Avebury, 136-144.

Holdsworth, D., 1994. Accountability: the obligation to lay oneself open to criticism. *In:* Chadwick, R.F. ed. *Ethics and the professions.* Aldershot, Avebury, 58-87.

HUGO Ethics Committee, 1996. *Statement on the principled conduct of genetic research.* Human Genome Organization HUGO, London. [http://www.gene.ucl.ac.uk/hugo/conduct.htm]

HUGO Ethics Committee, 2000. *Statement on benefit-sharing.* Human Genome Organization HUGO, London. [http://www.gene.ucl.ac.uk/hugo/benefit.html]

Illich, I., 1997. *Disabling professions.* Boyars, London.

Koehn, D., 1994. *The ground of professional ethics.* Routledge, London.

Langan, J.P., 1991. Professional paradigms. *In:* Pellegrino, E.D., Veatch, R.M. and Langan, J.P. eds. *Ethics, trust, and the professions: philosophical and cultural aspects.* Georgetown University Press, Washington DC, 221-235.

Pellegrino, E.D., 1991. Trust and distrust in professional ethics. *In:* Pellegrino, E.D., Veatch, R.M. and Langan, J.P. eds. *Ethics, trust, and the professions: philosophical and cultural aspects.* Georgetown University Press, Washington DC, 69-85.

Serageldin, I., 2002. The rice genome: world poverty and hunger - the challenge for science. *Science,* 296 (5565), 54-57.

Tur, R.H.S., 1994. Accountability and lawyers. *In:* Chadwick, R.F. ed. *Ethics and the professions.* Aldershot, Avebury, 58-87.

Veatch, R.M., 1991. Is trust of professional a coherent concept? *In:* Pellegrino, E.D., Veatch, R.M. and Langan, J.P. eds. *Ethics, trust, and the professions: philosophical and cultural aspects.* Georgetown University Press, Washington DC, 159-169.

Vyvyan, J., 1971. *The dark face of science.* Joseph, London.

Williams, B., 1996. *Freedom on probation: a case study of the Home Office enforced changes to the University education and training of probation officers (pamphlet).* Association of University Teachers.

12a

Bioscientists as ethical decision-makers

Matti Häyry[#]

The questions

I will address in this paper three questions. These are:
- Why should bioscientists be ethical decision-makers?
- What are bioscientists as ethical decision-makers?
- How can bioscientists act as ethical decision-makers?

Answers to these questions should clarify the role of geneticists, molecular biologists and other bioscientists in both private and public research organizations, and especially in their ethics committees.

Why should bioscientists be ethical decision-makers?

I can think of three separate sets of reasons why bioscientists should, in addition to their more familiar roles as researchers, administrators and entrepreneurs, also be prepared to pass ethical judgments concerning their own work and the work of their colleagues.

Because they are moral agents

The internal, autonomous, 'genuinely ethical' raison d'être for bioscientists to consider their work from this viewpoint is the moral agent's spontaneous need, or urge, to do the right thing, or to do things in the correct manner. If we go down this path, bioscientists can assess the ethical acceptability of their work from many different theoretical angles.

They can try to find out what the inner logic, or natural goal, of their activities is, and to estimate, to the best of their ability, how well what they do, or propose to do, follows this inner logic, or is likely to achieve the natural goal. In historical and philosophical terms, this is the Aristotelian, or 'teleological' approach to ethics. In research-ethics committee work, which is the most natural setting for the moral assessment of scientific work, this aspect is usually covered by the scientific evaluation of the plan's robustness in terms of methodology and the competence of the research team.

Alternatively, they can start from moral rules and principles, and consider what they do in the light of them. Historically and philosophically, this is the Kantian, or 'deontological' approach. In institutional ethics committees, this route is taken when members make appeals to the dignity, equality, autonomy, rights and vulnerability of the research subjects, or question the rational acceptability of the proposed work.

[#] Centre for Professional Ethics, University of Central Lancashire, Preston PR1 2HE, United Kingdom.
E-mail: mhayry@uclan.ac.uk

These are common themes in, for instance, stem-cell research and research into cloning humans.

Yet another 'internal' option is to assess proposals in terms of harms and benefits – to estimate how useful the performed or planned research can be in terms of human (or animal) welfare, and how harmful some of its consequences could be. The historico-philosophical roots of this approach are in eighteenth-century British utilitarianism, and it is commonly referred to as 'outcome-based ethics' or 'consequentialism'. This is partly covered in ethics-committee work by the scientific evaluation of the proposals, and partly by risk assessment and safety calculations.

Because they are professionals?

Another reason why bioscientists should be willing to judge their work ethically is a function of their role as experts in their field, and contenders to the useful label of being 'professionals'. Physicians and lawyers have traditionally benefited by claiming this label, and by formulating professional codes of conduct to back up this claim.

Professional codes should not be confused with the rules derived from autonomous ethics based on the moral instincts of individual moral agents. They exist for an external reason – namely, the need of the professionals to prove to the wider community that what they do is acceptable. Doctors stick needles into other people, they cut off body parts and expose others to near-lethal radiation, and they must have good grounds for doing these things and clear guidelines according to which they do them. Professional codes define, also to the professionals, but primarily to outsiders, the grounds for the privileged interventions and the limits within which they are acceptable.

Insofar as bioscientific research involves such interventions or deception or comparable atrocities, those conducting the research also need to prove their benevolence, trustworthiness and accountability to the wider community. This, in fact, is one of the original reasons for the establishment of ethics committees in hospitals, research institutes and universities. Externally they are, if you like, the window-dressing of research – the first thing you see when you look through the shop window. Internally, of course, they can be, or fail to be, the effective safety police they promise in their advert.

It is not self-evident that bioscientists can claim the status of being a unified professional group. I will return to this question in considering what bioscientists actually are as ethical decision-makers.

Because they are law-abiding citizens

Research-ethics committees, and other forms of institutional moral evaluation, increasingly owe their existence to national or regional legislation. This is the research-governance rationale for the establishment of such committees, and it is based on the political need to do things 'by the book', at least in publicly funded institutions.

The deeper justification for bioscientists evaluating each other's work is, in this line of thinking, essentially the nation's, or the international community's, commitment to democracy, transparency and accountability also in matters related to research. The idea is that individuals as moral agents cannot always be trusted to check dangerous activities if considerable personal gain is involved. The same applies to professional codes – if there are no public controls, how can the public be sure that researchers do not occasionally turn a blind eye to the activities of their peers?

Another way to formulate this reason for ethicality is to say that in jurisdictions where moral considerations are legally required it is prudential for bioscientists to engage in them. This interpretation would make their ethics entirely 'heteronomous', or externally initiated, as opposed to the 'autonomous' decision-making of Kantian moral agents.

What are bioscientists as ethical decision-makers?

Let me now turn to my second question. The three different reasons why bioscientists should become ethical decision-makers sketch three different pictures of them in this role.

Bioscientists as moral agents

It is unproblematic in any ethical theory that all competent human beings are moral agents. As such, they should make decisions based on the inner logic of things (the Aristotelian way), rationality (the Kantian alternative) or the consequences of actions (the utilitarian option).

It is not, however, easy to define what individuals should do when the different theories yield conflicting normative conclusions. Bioscientific activities have for the most part been criticized by (roughly) Aristotelian and Kantian ethicists, which indicates that prudential researchers would do well to assume the outcome-based model. But the advocates of the other approaches argue that this would be an immoral choice, and would leave many immaterial yet significant ethical issues unaddressed.

Furthermore, the choice of any moral theory exposes agents, at least potentially, to situations where their autonomous decisions go against the attitudes and ideals of legislators, ethicists or the general public. This does not necessarily show a flaw in the individual's thinking or actions, but it can make life difficult. The more idiosyncratic the chosen morality, the more problematic it can be to anchor one's ethics to one's individual moral agency.

Is bioscience a profession?

Bioscientists could avoid the problems of individual morality by assuming the role of professionals. But is bioscience a profession? I do not think that this question has been satisfactorily answered in the current discussion.

True professionals should, according to prevailing academic views, fulfil the following set of criteria legitimately to claim the status. They should have

- specialized knowledge;
- long and intensive academic studies;
- permanent careers;
- organization and self-rule within the group;
- as a group, a decisive role in the arrangement of the relevant studies and in the recruitment of new members to the group;
- a distinctive professional ethos or morality within the work; and
- positions of considerable responsibility in communities and societies.

When all these conditions are met, members of a group can confidently call themselves professionals. But when some of them are not met, the situation is more complicated.

Bioscientists fulfil many of the criteria, but some focal questions remain. Do they have organization and self-rule within themselves? Do they have a decisive role in the arrangement of studies and recruitment? And do they have a distinctive ethos?

Chapter 12a

The ethos of a profession is, in the case of physicians and lawyers, expressed in a code of conduct which is accepted and endorsed by the members of the group. I have not seen good ethical codes formulated by bioscientists, and I am not even convinced that 'bioscientists' as a group for itself exists. Medical researchers often argue that the codes of physicians should be used as a model for bioscientific codes. But not all geneticists and molecular biologists want to be identified with this existing profession.

Bioscientists as legal positivists
The problems inherent in the application of individual and professional ethics have led some bioscientists to believe that laws are the only possible source of norms in their field. This line of thinking is quite natural in societies where people do not tend to question the legitimacy of prevailing regulations. The idea, sometimes identified with the doctrine of legal positivism, is that the laws of the country are justified because they are the laws of the country.

The growing impact of international regulation has, however, changed the situation, at least in Europe. Citizens of the member states of the European Union do not always find the ethical, and often religious, ideologies of other states acceptable. As a corollary, even bioscientists who have ungrudgingly acceded to national restrictions in their own field have begun to have doubts concerning imported rules. And when choices have to be made between competing ideas about regulation, ethics in the other two senses can once again be seen as the answer.

How can bioscientists act as ethical decision-makers?

Although it is, in theory, hard to define bioscientists as ethical decision makers, it can still be possible to outline what they could do to ensure the moral acceptability of their work.

Sensitivity and competence
A standard line in the terms of reference of research-ethics committees is that they should safeguard the health and welfare of persons and the environment, and to see to it that unnecessary harm is avoided, the autonomy and dignity of those involved protected, and the requirements of justice promoted in the process. How should all this be done in the assessment of research plans or ongoing projects? One way would be to employ the following generic formula:

$$e = mc^2.$$

In this formula, as I use it here, the symbols stand for the following things:

e = ethical acceptability of proposed research
m = moral non-sensitivity of proposed research
c = competence of research plan and research team.

(I am fully aware that the symbol 'c' cannot rightly be used in two meanings in this equation, and that the correct formulation would be $e = mc_{rp}c_{rt}$, where c_{rp} and c_{rt} represent the two types of competence. I could not, however, resist the temptation of using the shape of the better known law)

How is this play with symbols useful in our present context? Well, it identifies the three main criteria for assessing research, or at least research proposals – the sensitivity of the topic, the quality of the plan and the professionalism of the team. It also tells us, because only straightforward multiplication is involved, that total failure in any area of assessment lowers the acceptability considerably and should probably lead to the rejection of the proposal. If any one element in the multiplication is given the value zero, then this value is transferred to its ethical acceptability as well. And it tells us that partial failure in any one area can, to a certain extent, be compensated by excellence in other areas. (Due to this quality, the formula could be called the 'general law of ethical relativity')

An important question in any attempt to isolate and then recombine the aspects of complex phenomena is, 'Can the model be quantified?' In this case, tentatively, perhaps. We could say, for instance, that:
- if the proposed research is totally non-sensitive, its value is 1;
- if it is highly, but not impossibly, sensitive, its value can be 0.5; and
- if it is too sensitive to be conducted, its value approaches 0.

And we could say that:
- if the proposal or the research team is extremely competent, the value of 'c' is 2;
- if these are adequately competent, the value could be 1; and
- if they are totally incompetent, the value would be 0.

These figures are, of course, totally arbitrary, but they produce some interesting equations.

The value of 'e', or the ethical acceptability of the research in question, would with these figures range from 0 to 4. If we then stipulate that 1 is the minimum needed for ethical approval, we can reach this result in several ways:

$$e = 1 = 1 \times 1 \times 1.$$

This says that with a non-sensitive topic, a merely 'adequate' level of competence both in the plan and the team would be acceptable.

$$e = 1 = 0.5 \times 2 \times 1 = 0.5 \times 1 \times 2.$$

This indicates that a relatively high sensitivity of the topic can be compensated either by good planning or a good team history in this type of research.

$$e = 1 = 0.25 \times 2 \times 2.$$

This expresses two important ideas. The first is that extreme competence goes a long way in dealing with extremely sensitive topics. The second is that there are levels of sensitivity beyond which compensation is simply impossible.

If all this can be agreed upon in one way or another, all we would need in order to complete the model would be an operationalization of the key concepts – 'sensitivity', 'plan competence' and 'team competence'. Ideally, this task could be performed in co-operation between ethicists, who should find out how health, welfare, dignity, autonomy and justice are to be balanced in the measurement of sensitivity; and scientists, who should define the costs and benefits of the proposed work, and the technical competence of the plan and the researcher.

Since all these definitions and weighings are complicated, and partly controversial, I would not hold my breath waiting for a general theory of ethical acceptability to emerge from this model. But perhaps it is possible to go one step further in this direction by focusing on the professional-code aspect of ethics.

Chapter 12a

Principles and values

Professional codes often include a set of background principles, or values. Judging by contemporary ethical debates, bioscientists, if they can unite as a profession, can choose from two main approaches as regards such principles or values. It has been argued that the norms presented by philosophers and theologians in these debates are too far removed from practical concerns to be of any real use to scientists. I do not necessarily disagree. But let me briefly present two models of bioethics which seem to be popular in public discussions, and say a few words about their implications on bioscience.

The first is the 'four-principles approach', mainly advocated by American bioethicists. It states that an ethically good choice, or course of action, fulfils the following criteria:

- It respects the autonomy of individuals.
- It does not inflict unnecessary harm on anybody.
- It does some good to somebody.
- It does not violate any precepts of justice.

Some critics have argued that this American approach reduces morality to consumer choices by overemphasizing the autonomy and material well-being of independent and affluent individuals. Depending on the interpretation of the principles this concern is partly valid. At least in the context of biosciences, it is true that *if* autonomy means market freedom, *if* harms and benefits are balanced in expert risk assessments, and *if* justice means giving the researchers their dues, *then* this model can give scientists a licence to do almost anything, provided that they do not harm any identifiable individuals by doing so. But this is not the only feasible interpretation of the central concepts.

Many European ethicists have tried to find fundamental moral principles in other directions. Their favourite criteria for a sound ethical decision seem to include:

- It does not violate the dignity of human (or other living) beings.
- It is precautionary, that is, it takes into account all conceivable consequences, including those which are presently unforeseeable.
- It respects solidarity, that is, it does not benefit one group of people at the expense of others.

The most prominent objection to the use of these principles in public debates is that they are vague, and can therefore be employed to support bans on anything the ethicists employing them do not understand or like. Certainly many voices against the development of biosciences have been heard from the advocates of this set of rules. Again, however, the values of dignity, precaution, and solidarity can also be interpreted in other ways.

Although these two groups of principles can be given different meanings, knowledge of them could be useful to bioscientists, if and when they set out to formulate a unified professional code for themselves. The picture I have given here can be given more colour and content by bioethical studies, and the reflection of contemporary principles and values would probably serve the emerging professionals better than attempts to invent rules from scratch.

Law or morality?

Laws regulating bioscientific research have, to a certain extent, been arrived at by examining the nature of the work professionals in this field do, and by addressing concerns encapsulated in the principles of 'autonomy', 'non-maleficence', 'beneficence', 'justice', dignity', 'precaution' and 'solidarity'.

But if the role of bioscientists as ethical decision-makers is to be taken seriously, they should, as individuals or as groups, scrutinize their own work and values, and present to others a code, or a model, by which they themselves would like their practices to be assessed. This would serve two ends: they would give legislators and the general public a better 'expert' view of what their activities are all about, and they would have the opportunity to define their own values as a group.

References

Some of the ideas discussed in this paper have been further examined in:

Häyry, M., 2003a. Do bioscientists need professional ethics? *In:* Häyry, M. and Takala, T. eds. *Scratching the surface of bioethics*. Rodopi, Amsterdam, 91-97.
Häyry, M., 2003b. European values in bioethics: why, what, and how to be used? *Theoretical Medicine and Bioethics,* 24 (3), 199-214.

12b

Comments on Häyry: Assessing bioscientific work from a moral point of view

Robert Heeger[#]

A difficulty in making moral assessments

There are *good reasons* for bioscientists to assess their work from a moral point of view. In his paper, Matti Häyry mentions three important reasons. If bioscientists are a profession and assess their work by a professional code, they can prove to the wider community that they are benevolent, trustworthy and accountable. If national legislation or supranational directives require moral assessments of their activity, it is prudent for bioscientists to engage in such assessments. But first and foremost bioscientists are competent moral actors who feel the spontaneous need to do the right thing. This reason is the most important one, because it is, unlike the other two, not primarily external or prudential but both internal and genuinely moral.

Matti Häyry's *main question* is how bioscientists could perform the task of assessing their work from a moral point of view. He brings to the fore that they could be guided by different ethical approaches. They could join a teleological, Aristotelian, theory according to which their projects are to be measured against the inner logic or the goal of their activity; or they could follow a deontological, Kantian, theory which requires them to consider whether their projects are compatible with principles of respect for the dignity, equality or autonomy of research subjects; or they could assent to a consequentialist, utilitarian, theory which says that the moral acceptability of their projects depends solely on the beneficial or harmful effects on all those concerned.

The diversity of these approaches causes a *difficulty* in making moral assessments: it is not easy to define the right thing to do when the different theories yield conflicting normative conclusions. To try to evade this difficulty by opting for one of the theories while disregarding the other two would not lead to a lasting solution, because one would be liable to criticism for leaving significant ethical issues unaddressed. This would for instance apply to opting for a consequentialist theory. One would then be liable to an Aristotelian or a Kantian criticism.

A proposed forward movement

How could the above-mentioned difficulty be overcome? Häyry presents an outline of what bioscientists could do to ensure the moral acceptability of research plans or ongoing projects. His proposal consists of two parts. First, he suggests assessing the moral acceptability of a research plan or project by using *three criteria*, viz. the moral sensitivity of the topic, the quality of the research plan, and the professionalism of the team. One could try to quantify each of these features in the following manner: the

---

[#] Centre for Bioethics and Health Law and Faculty of Theology, Utrecht University, Heidelberglaan 2, 3584 CS Utrecht, The Netherlands. E-mail: RHeeger@theo.uu.nl

higher the quality of the research plan and the higher the professionalism of the team, the higher the quantitative values assigned to them, respectively, but the higher the moral sensitivity of the topic, the lower the quantitative value, approaching zero if the proposed research is too sensitive to be conducted. The moral acceptability of the research in hand could then be decided on the basis of the product of the three values. The product should be at least as large as the minimum requirements for moral acceptability stipulated in advance.

The second part of the proposal concerns the *definition of the key concepts*, viz. 'moral sensitivity', 'quality of the research plan' and 'professionalism of the team'. To define these concepts is a very difficult task. Take for example the first concept, moral sensitivity. This concept has several constituent parts, such as health, welfare, dignity, autonomy and justice. In order to make the concept suited for the suggested moral assessment, the content of these parts should be clarified. Moreover, it should be found out how the parts are to be balanced in the measurement of moral sensitivity. Noticing both kinds of problems Häyry makes a distinction between what ideally *should* be done and what actually *can* be done. Generally acknowledged definitions of the key concepts may, at least at present, be beyond reach, but one could perhaps take a step towards such definitions by focusing on a professional code. There are several codes which include ethical background principles, viz. principles of respect for autonomy, of non-maleficence, beneficence, justice, dignity, precaution and solidarity. Bioscientists could make use of them, if and when they formulate a professional code for themselves.

The guiding role of principles

Häyry's proposal has at least two appealing traits. First, he is not content with outlining three criteria for moral acceptability but also cares about the definition of the concepts in question. Second, he does not capitulate in the face of the complexity of the concepts but suggests focusing on ethical principles as a step in the direction of the key concepts. For these reasons the proposal merits assent. However, it also gives rise to a critical question regarding the problem-solving role that it expects ethical principles to play.

Suppose we try to assess the moral acceptability of a project by means of the above principles. How can we avoid being driven back to a difficulty similar to that mentioned in the beginning? Put in other words, how can we prevent finding ourselves in a situation of indecision, this time not on the level of ethical theories but on the level of ethical principles? After all, several principles may be relevant in assessing the project, and these principles may yield conflicting normative conclusions.

One step in answering this question is that we should become clear about the *meaning* of the different principles. To mention principles is not enough. If we are to use them in order to assess a project, we must be clear about their content. An ethical principle is not a formula or an authoritative prescription which we simply have to comply with. Rather, it presupposes substantive moral judgement. Take the first principle mentioned above, respect for autonomy. Suppose it is to apply to a medical researcher setting out to perform a clinical trial. For the medical researcher to know what respect for autonomy means, it is not enough that she knows the lexical meaning of the words 'respect' and 'autonomy'. She must also know situations in which respect for patients is required, she must appreciate the value of autonomy, and she must appreciate the moral relevance of this value.

Another step is realizing that the *application* of principles requires subsidiary considerations of two kinds. One kind can be called considerations of application. Such considerations are closely connected with the meaning of a principle. They regard subsidiary questions as to what the respective principle exactly requires and whether it really is applicable to the case in hand. The other kind can be called considerations of specification. These considerations are important when several relevant principles conflict, that is, cannot be jointly satisfied. Take the second and third principle mentioned above, non-maleficence and beneficence. Suppose both are applicable to an intended animal experiment, because the purpose of the experiment is clearly connected with beneficence and the infliction of suffering on the animals is a case of maleficence. Considerations of specification take into account that both beneficence and inflicting suffering, that is violating the principle of non-maleficence, come in degrees. Relevant questions are for example: How important is the objective of the experiment, how grave is the infliction of suffering, is a reduction of suffering or a replacement of the experiment possible? Such considerations help us when we cannot fully comply with both principles but have to work out a package of actions that meets the principles only to the largest possible extent. Also in this arbitration between the conflicting claims of principles moral judgment is necessary.

So, ethical principles of a professional code for bioscientists can play a guiding role only if they are firmly rooted in the professionals' capacity for moral judgment.

References

Beauchamp, T.L. and Childress, J.F., 1979. *Principles of biomedical ethics*. Oxford University Press, New York.
Beauchamp, T.L. and Childress, J.F., 1983. *Principles of biomedical ethics*. 2nd edn. Oxford University Press, New York.
Beauchamp, T.L. and Childress, J.F., 1989. *Principles of biomedical ethics*. 3rd edn. Oxford University Press, New York.
Beauchamp, T.L. and Childress, J.F., 1994. *Principles of biomedical ethics*. 4th edn. Oxford University Press, New York.
Den Hartogh, G.A., 1999. General and particular considerations in applied ethics. *In:* Musschenga, A.W. and Van der Steen, W.J. eds. *Reasoning in ethics and law: the role of theory, principles and facts*. Aldershot, Ashgate, 19-47.
Richardson, H.S., 1990. Specifying norms as a way to resolve concrete ethical problems. *Philosophy and Public Affairs,* 19 (4), 279-310.

NEW DEVELOPMENTS

13

The human genome: common resource but not common heritage

David B. Resnik[#]

Introduction

Since the 1980s, biotechnology and pharmaceutical companies have aggressively pursued intellectual property rights in biological materials in order to protect their proprietary interests and secure a reasonable return on their research and development costs. Although the biotechnology industry is still only in its infancy, it has generated billions of dollars in private investment, hundreds of thousands of jobs, as well as the promise of new treatments for various diseases and substantial improvements in agricultural production. It has also created a storm of ethical and political controversy. Many new applications of bioscience, ranging from gene therapy and pharmacogenomics to genetically modified foods and animals, require the ability to isolate, purify, analyse, clone and modify DNA. It should come as no surprise, then, that the various stakeholders in biotechnology, including private companies, universities and government agencies, have sought to acquire intellectual property rights in DNA. It should also come as no surprise that ethical and political controversies have erupted from the intellectual property race in biotechnology.

Those who oppose proprietary control of DNA have voiced a variety of objections to the patenting of DNA sequences, including the claim that patenting DNA violates human dignity, the assertion that patenting DNA violates the sacredness of nature, and the hypothesis that patenting DNA will have adverse effects on the progress of science, medicine and agriculture (for further discussion, see Resnik 2003). This essay will not attempt to explore all of these different objections to DNA patenting but will focus on one particular objection that has had considerable international influence, the idea that the human genome is the common heritage of mankind (referred to hereinafter as the 'common heritage' idea).

The common-heritage idea has influenced ethical and policy debates concerning the commercialization of the human genome. Many different organizations have championed this idea, including the Human Genome Organization (HUGO) Ethics Committee (2000), the Council on Responsible Genetics (CRG 2000), the International Federation of Gynaecology and Obstetrics (1997), The Parliamentary Assembly of the International Council of Europe (Council of Europe 2001) and the United Nations Educational, Scientific and Cultural Organization (UNESCO 1997). A UNESCO declaration states that, "The human genome underlies that fundamental unity of all members of the human family...in a symbolic sense, it (the human genome) is the heritage of humanity...The human genome in its natural state shall not give rise to financial gains" (UNESCO 1997). Additionally, some scholars, such as

---

[#] Department of Medical Humanities, The Brody School of Medicine, East Carolina University, 2S-17 Brody Building, Greenville, NC, 27858, USA. E-mail: resnikd@mail.ecu.edu

Looney (1994) and Sturges (1997) have argued that the human genome should be viewed as our common heritage, while others, such as Juengst (1998), Ossario (1998), Spectar (2001) and the Nuffield Council on Bioethics (2002) have critiqued this idea.

The claim that the human genome is our common heritage coincides with the debate about patenting of DNA sequences that began in the 1990s. People opposed to DNA patenting argued the common-heritage idea has important policy implications for the commercialization of human DNA. Some writers argued that viewing the human genome as our common heritage implies that there should be no patents on human DNA sequences (CRG 2000). This paper will examine and critique the idea that the human genome is the common heritage of mankind. It will argue that the human genome is not literally our common heritage; it is best viewed as a common resource, but not as our common heritage. Since the genome is a common resource, the patenting of DNA is morally acceptable, provided that we honour out moral duties to the genome, which include duties of stewardship and justice. This essay will give a brief overview of treating DNA as intellectual property before proceeding to the main arguments.

Patent Law and DNA

To understand how one can patent a DNA sequence, it will be useful to review quickly U.S. patent law. European patent law is similar to U.S. law in many respects (Nuffield Council on Bioethics 2002). A patent is a right granted by the government to exclude others from using, making or commercializing an invention for a limited period of time. In the U.S., the life of a patent is 20 years from the date of the application (Miller and Davis 2000). The legal basis for patents has its roots in the U.S. Constitution, which states that Congress shall have the power "To promote Progress of Science and useful Arts, by securing for limited Times to Authors and Inventors the exclusive right to their respective Writing and Discoveries" (*United States Constitution*). In 1790 the U.S. enacted a federal law, the Patent Act (*Patent Act 35 USC 101* 1952, 1995), to implement this constitutional mandate. The Patent Act has been amended several times (Miller and Davis 2000).

The main ethical and policy rationale for granting patents is utilitarian: patents promote scientific and technological progress by giving financial incentives to inventors, investors and entrepreneurs (Resnik 2001b). Scientific and technological progress are valuable for their own sake and because they contribute to economic growth and to advancements in medicine, engineering and agriculture. One reason why people invest time and money in developing inventions is that they expect to be able to make money from those inventions. Prior to the development of the patent system, inventors and craftsmen would use trade secrecy to protect their intellectual property. The patent system encourages inventors to forego trade secrecy and make their inventions available to the public. Under a theory known as the patent 'bargain', the government grants an inventor a private right in exchange for public disclosure of information in the patent application (Miller and Davis 2000).

Before an invention can be patented, it must qualify as a patentable subject matter. Under U.S. law and European law, one may obtain a patent on any new and useful process, product or improvement on a process or product (Miller and Davis 2000). For example, a light bulb would qualify as a product; a method for manufacturing light bulbs would qualify as a process, and an energy-saving light bulb might qualify as an improvement on a product. In biotechnology, one can patent various biochemical products, such as DNA sequences, as well as biochemical processes, such as methods for isolating, purifying, cloning, modifying, analysing and manufacturing DNA (Eisenberg 1990).

In applying patent law to particular items, the courts have drawn distinctions between products of nature, which are not patentable, and products of human ingenuity, which are patentable (Miller and Davis 2000). For example, the courts have held that laws of nature, natural phenomena and naturally occurring species are not patentable, because they are products of nature. However, a genetically engineered plant or animal can be patented because it is a product of human ingenuity (Diamond vs. Chakrabarty 1980).

Although these distinctions relating to subject matter have a philosophical tone, they are best understood as pragmatic exercises in line-drawing: these distinctions are based on political and public-policy concerns rather than on any objective, metaphysical theory that divides the world into 'products of nature' and 'products of human ingenuity' (Resnik 2002).

The patenting of DNA sequences has posed a conceptual challenge for patenting agencies because DNA occurs in a natural state in organisms. How can DNA be a product of human ingenuity? To deal with this problem, patenting agencies have held that DNA is similar to other chemicals found in nature that can be patented under the doctrine of isolation and purification, such as vitamin B12 or human growth hormone (Doll 1998). By isolating DNA from its natural state and reproducing the compound in a highly purified form, scientists have used a sufficient modicum of human ingenuity to transform DNA into a patentable invention. Someone who has patented human DNA does not own a human being or even have patent rights over a living human being; he only has patent rights over some of their DNA produced under laboratory conditions (Resnik 2001a).

In order to obtain a patent an inventor must submit an application to the patent agency that describes the invention in sufficient detail to allow a person trained in the relevant practical art to make and use the invention. Once the patent is awarded, the application becomes a public record. To receive a patent, the invention must be novel (it has not been previously invented or disclosed in the prior art), non-obvious (it is not obvious to someone trained in the relevant practical art), and useful (the invention serves some practical use) (Miller and Davis 2000). Once an inventor obtains a patent, he (or she) may assign the patent to a university or corporation, or he may license others to make, use or commercialize the invention. Under U.S. law, the inventor is also free to do nothing with the invention and keep it off the market. Unlike Europe, the U.S. has no compulsory licensing provision in its patent law (Miller and Davis 2000).

If someone makes, uses or commercializes his invention without the patent holder's permission, the holder may sue that person for patent infringement. The U.S. courts have recognized (but rarely used) an exemption to patent infringement know as the research exemption. The research exemption allows a person to use or make an invention for purely 'philosophical' research that has no prospect of any commercial application (Karp 1991). Since almost all research in biotechnology has potential commercial applications, the research exemption may not be available to most university-based researchers (Resnik 2001b). Some writers have suggested that the research exemption should be clarified and legislatively reinforced in order to promote progress in biotechnology and biomedicine (Nuffield Council on Bioethics 2002).

During the term of the patent, patent holders have exclusive rights pertaining to their inventions. They derive economic benefits from their patent during its lifetime, such as sales of the product or service and licensing. Patents are generally far more lucrative than copyrights in biotechnology, although copyrights on databases could hold considerable financial promise (Resnik 2003). Trade secrecy is not a very attractive form of intellectual property protection because it is very difficult to keep secrets in biotechnology. Unless a researcher invents an entirely new product or process, i.e. one with no simulacrum in the natural world, then other people will be able to discover his product or process simply by reproducing it from available natural materials and phenomena (Resnik 2003).

The common-heritage idea

Having set the legal and ethical context for the common-heritage idea, we can explore the argument in more detail and critique it. As noted earlier, the common-heritage idea asserts that the human genome is the common heritage of humanity. How should we interpret this idea? What does it mean to say that the human genome is the common heritage of humankind? A heritage is usually defined as a property that can be inherited or passed down from one generation to the next (*The American Heritage® dictionary of the English language* 2000). To say that something is a common heritage, there must be a) an identifiable thing (or set of things) that is (are) inherited; and b) an identifiable person (or set of persons) that inherit(s) the heritage; c) an identifiable person (or group of people) who bequeath(s) the inheritance. For example, suppose a man dies and leaves some land to his four children. Each child has 25 acres of land and access to a river that runs through each child's land. Under these conditions, each child has a personal heritage, i.e. his or her land, as well as a common heritage, the river. The man bequeaths the river and the land.

If the human genome were literally mankind's common heritage, then DNA patenting would be, for all practical purposes, illegal, because one would need to obtain consent from every human being to commercialize the human genome, since every human being would have a property interest in the genome. In the river example, no child should be able to commercialize the river without obtaining consent from the other children, because they all have a property interest in the river. The Law of the Sea Convention, adopted by the United Nations, makes explicit use of the common-heritage idea (Sturges 1997). Under this doctrine, no country can appropriate for itself the territories held in common, such as the moon, Antarctica or the deep sea beds.

Some scholars and organizations have argued against any DNA patenting on the grounds that the human genome is literally mankind's common heritage. There are at least two reasons why one might regard the human genome as our common heritage. First, we all share a common ancestry through the genome. Although different human populations have evolved somewhat since the origins of *Homo sapiens* over 1 million years ago, every human being can trace his or her ancestry back to the founding members of our species. Second, human beings have almost all of their genes in common: we share over 99% of our genes.

A moment's reflection on the nature of DNA is sufficient to show that there are some significant problems with regarding the human genome as mankind's common heritage. The first problem is that there is not a single, identifiable thing (or set of things) that constitute(s) the human genome. There is a significant amount of genetic variation among members of the species *Homo sapiens*. Although human beings share most of their DNA, there are thousands of single-nucleotide polymorphisms (SNPs), which vary from person to person (Venter et al. 2001). Human beings also exhibit a great deal of variation in haplotypes (or patterns of sequence variation). The second problem is that there is not a single, identifiable set of people who inherit the human genome. Human beings share 98.5% of the DNA with chimpanzees, 95% with other primates, a great percentage of their DNA with other species, including fruit flies and yeast (Venter et al. 2001). So, only 1.5% of the human genome is actually 'our' common heritage; the other 98.5% of the genome is the heritage of other species. Should we say that the human genome is also the common heritage of the chimpanzees, the primates, all mammals, or even yeast? Does it make sense to say that non-human species can have property interests? The third problem is that we cannot identify the persons of set of persons who have bequeathed our DNA to us. Did our ancestors ever intend to bequeath their DNA to all of humanity? These three problems show that is does not make much sense to regard the human genome as literally our common heritage. The common heritage idea may have symbolic importance, but it is an empirical fiction (Juengst 1998).

If we do not regard the human genome literally as humankind's common heritage, we could still view it is symbolically humankind's common heritage. The UNESCO declaration speaks of the human genome as our common heritage in a symbolic sense (UNESCO 1997). Rejecting the literal interpretation of 'common heritage' in favour of the symbolic interpretation has important implications for ethics, law and public policy. Since the human genome is not literally our common heritage, patenting human DNA is not *ipso facto* immoral or illegal. The morality and legality of patenting depends on the facts relating to the type of patenting in question as well as the values at stake. Some types of patenting may be immoral, some may be illegal, and some may be both immoral and illegal. We have to examine each type of patenting on its own merits to determine its morality and legality.

The human genome as a common resource

Suppose that we think of the human genome not as humankind's common heritage but as a common resource. What follows from this postulate? First, the common-resource idea does not imply that every person has an ownership interest in the genome; it does not create a common property right in the genome (Ossario 1998). Individuals, corporations or countries may commercialize the genome without obtaining permission from every human being. Second, the common-resource idea does not imply an 'anything goes' approach to our duties toward the human genome, since we have moral duties relating to common resources. It is morally acceptable to commercialize the Earth's resources, provided that we honour our moral obligations vis-à-vis those resources. We have duties to take care of these resources and use them wisely and fairly. Likewise, we have moral duties relating to the human genome as a common resource, even though we may commercialize this resource.

If we think of the human genome as a common resource, we can apply some insights from environment ethics to genome policy. Duties to the environment include duties of stewardship and justice, which are based on duties to current and future generations (Rolston 1994). We should take care of the oceans, for example, so that people will be able to use and enjoy the oceans both now and in the future. Similar duties also apply to our treatment of the human genome. If we think of the genome in this way, then the duties of stewardship and justice arise from the fact that current and future generations have a common interest in the human genome, even if that interest is not a property interest.

Duties of stewardship

If something is a common resource, we have duties of stewardship toward that resource. A steward is someone who is in charge of taking care of something for someone else. Like a trustee, a steward has duties to preserve and develop the thing he or she is entrusted with. In a sense, we are entrusted with the human genome in the same way that we are entrusted with the earth, or an investment banker is entrusted with an investment portfolio. Our duties of stewardship toward the human genome should include protecting the genome from harm, such as loss of genetic diversity or the propagation of harmful (human-induced) mutations.

Some writers, such as Juengst (1998), have expressed some concern about the eugenics implications of the stewardship idea: if we have an obligation to avoid harming the genome, don't we also have an obligation to benefit the genome by eliminating 'undesirable' mutations? The trouble with the idea of 'benefiting' the genome is that it could be used to justify the horrors associated with the eugenics movements in the 20th century, including Nazism. There is a slippery slope from attempting to improve the genome to attempting to purify the human race, as well as a slippery slope from attempting to prevent genetic harms to seeking genetic perfection.

Clearly, one needs to describe carefully the duties of stewardship of the genome to avoid eugenics implications. Certainly, we should not engage in forced sterilization, restricted procreation, ethnic cleansing, genocide, genetic discrimination or other immoral activities under the mistaken idea that we should purify the genome. On the other hand, most people would agree that we have obligations not to engage in activities such as cloning and germ-line manipulation, if we determine that these activities pose a significant risk to future generations as well as a threat to the human gene pool. We must find some way of drawing a distinction between the obligation to avoid harming the human genome and the obligation to benefit the human genome. Although stewards normally have positive duties to benefit those things that they are entrusted with, there are sound moral reasons that these positive duties of stewardship should not extend to the human genome until we have a better understanding of the difference between therapy and enhancement in human genetics (for further discussion, see Buchanan et al. 2000).

Duties of justice

If something is a common resource, we also have duties to use the resource justly and fairly. We have duties relating to the sharing of benefits derived from the resource. Current generations should share the resource with each other and with future generations. While most people will agree that we have some duties relating to the sharing of the benefits from resources, few people will agree on the precise way in which benefits should be shared, because benefit sharing raises fundamental problems concerning distributive justice. Distributive justice addresses questions of how we should distribute benefits and burdens in society (Rawls 1971). Problems relating to distributive justice are some of the most contentious issues in contemporary moral and political philosophy. There currently is no consensus among scholars, commentators, politicians or the public concerning the substantive principles of distributive justice, even though there is a widespread agreement that considerations relating to justice are important in public policy debates. In response to these disagreements about substantive principles of justice, many writers have urged that we should develop theories of justice that focus on procedural notions of justice and fair procedures (Rawls 1993; Gutmann and Thompson 1996). Since distributive justice is a very complex and controversial topic, there is not adequate space in this essay to discuss the strengths and weaknesses of all the various theories, concepts and principles of justice. I will therefore limit my discussion to theories, concepts and principles of justice that have special relevance to benefit-sharing issues in human DNA patenting.

To begin the discussion of benefit sharing in genetic research, let's consider the infamous case of John Moore. Even though this case does not involve a DNA patent, it merits discussion because it illustrates some potential local benefit-sharing problems that can arise in DNA patenting. Moore contracted hairy-cell leukaemia, a rare form of cancer, in 1976. Dr. David Golde, Moore's physician at the University of California, Los Angeles (UCLA) Medical Center, recommended that Moore undergo a splenectomy. After Moore's spleen was removed, Golde asked Moore to make several visits to the Medical Center, so that Golde could take some additional samples of Moore's blood, skin, bone marrow and sperm. Golde lied to Moore and told him that these samples were needed to monitor his health. In reality, he used these extra samples to develop a cell line from Moore's tissue. Moore's tissue had a great deal of potential commercial value because it was overproducing lymphokines, which are proteins that play a key role in the immune system. The market for these compounds was estimated to be $1 to $4 billion. Golde and his research assistant signed agreements with the University of California and several pharmaceutical companies to develop the cell line. They also applied for and obtained patents on the cell line, which they assigned to the University of California. Moore eventually found out that he had been deceived, and he sued Gold, his assistant, the private companies and the University for medical malpractice and for conversion, i.e. substantial interference with personal property. The case eventually reached the Californian Supreme Court, which ruled that Moore did not have property interests in the cell line and could therefore not prove the tort of conversion. The researchers had property interests in the cell line because they had gone to the trouble of isolating, purifying and culturing the cell line. The cell line was their invention, and they had property interests in the cell line as patent holders. In the end, a divided court acknowledged that the defendants were negligent because they failed to obtain adequate informed consent, but it did not grant any property rights to Moore (Moore vs. Regents of the University of California 1990).

Although the Moore case involved a patent on a cell line, it could just as easily have involved a patent on a human gene. Indeed, a patent on a gene that codes for lymphokines might be even more valuable than the special cell line. There are many ethical problems with the Moore case, including deception, manipulation, fraud and inadequate informed consent. Although the court did not find that Moore had a property interest in his own cells, one does not need to make this assumption in order to assert that the researchers, the company and the university had a moral duty to share benefits with Moore and that they violated that duty. Moore provided the cells that became their gold mine. Although he did not deserve to be listed as a co-inventor on the invention, he made an important contribution to the invention. Without him, there would have been no invention. Thus, a principle of sharing benefits based on contribution would support sharing benefits with Moore, depending on the significance of his contribution. Of course, many different parties contributed to the invention. The inventors contributed labour, effort, skill and knowledge. The company contributed money. The university contributed its facilities, laboratories, technical support and supplies. Depending on how one measures these other contributions, Moore's contribution may have amounted to only 1% of the total. However, even if the invention netted $100 million in profit, a 1% share would still be worth $1 million. The bottom line in this case is that Moore got nothing, which is unconscionable.

For a different benefit-sharing case, consider the patent on the Canavan gene. Mr. and Mrs. Daniel Greenberg had two children who were born with Canavan disease, a rare neurological disorder that occurs almost exclusively in Ashkenazi Jews. The Greenberg's first child died when he was 11 years old. Their second child also developed the disease. The Greenbergs led an effort to identify the mutation that causes Canavan disease, and they enlisted the assistance of Dr. Reuben Matalon, a physician who was working at the University of Illinois Hospital in Chicago. The Greenbergs helped Matalon acquire skin, blood and urine samples from diseased children and their parents. They also raised about $100,000 in money to support the project. Miami Children's Hospital (MCH) soon hired Matalon to establish a centre for research on genetic diseases, and spent $1 million per year in support of his research. Matalon isolated the gene that causes Canavan disease in 1993. MCH applied for a patent on the gene, which the Patent and Trademark Office awarded on October 21, 1997. Matalon assigned all of his patent rights to MCH (Kolata 2000).

After MCH had obtained rights to the patent, it decided to charge royalties of $12.50 per test to laboratories that perform the test. The hospital planned to use the money from these fees to help offset the costs of research and development and publicity. MCH considered $12.50 to be a nominal and very reasonable royalty fee for the test. By comparison, Myriad Genetics has charged up to $1200 in licensing fees for its BRCA1 and BRCA2 tests (Foubister 2000). People from the Canavan community objected to the $12.50 licensing fee, however. They argued that MCH should make the test available to the public and that laboratories should be able to perform the test without paying any licensing fees. The Greenbergs and several other parties filed a lawsuit against MCH and Matalon in a Chicago federal court, alleging breach of informed consent, fraud, unjust enrichment, conversion and misappropriation of trade secrets. Recently, a federal court in Miami dismissed all of these claims except the unjust enrichment claim. The court found that the plaintiffs had invested enough money in the Canavan research that they could go forward with a claim of unjust enrichment against MHC (Greenberg v. Miami Children's Hospital Research Institute 2003).

The defendants in the Canavan case do not appear to be as unethical as the defendants in the Moore case. First, it does not appear that MHC and Matalon deceived people who contributed DNA samples to the research project. They did not take these samples in secrecy or under manipulative conditions. Second, the profit motive was probably not a major factor in the decision to charge licensing fees for the test, since $12.50 for a license is a very nominal fee. By comparison, a license for Microsoft Windows® software costs about $200. Unlike the Moore case, it does not appear that MHC will gain billions of dollars from its patent.

On the other hand, MHC, like the defendants in Moore, failed to establish a plan to share benefits with the Canavan community. It did not develop a plan to give members of the community money, healthcare, education or some other benefit as compensation for their contributions to the research. It is also did not consult with members of the community or patients about how benefits would be shared. If one accepts the principle that benefits should be shared based on contributions, then one might argue that the MHC failed to share benefits with the Canavan community, who deserved some form of compensation. Although other parties contributed time, efforts, skills, knowledge, facilities, technical support, supplies and money, members of the Canavan community contributed essential resources. Without their tissue donations, there would be no genetic test. MHC might reply, however, that it has already compensated the community for its contribution by developing the test. The test will benefit couples that carry the disease and allow them to prevent the birth of children with this crippling and painful illness. It has shared benefits from the community.

This reply raises an important point: what is just (fair or equitable) benefit sharing in genomics research? How much of a benefit should subjects and communities receive for their participation? Should researchers and companies provide them with financial compensation for their participation or with some other type of benefit, such as education or healthcare? Many organizations and scholars agree that there should be some type of benefit sharing in the commercialization of the human genome (Knoppers 2000; Human Genome Organization Ethics Committee 2000). The really hard questions have to do with the precise details concerning the structure of benefit sharing in any particular case.

Although developing a test or treatment is often a legitimate form of compensation for one's contributions to biomedical research, sometimes it may not be adequate. In this case, since the Canavan test is likely to be not very profitable, due to the small patient population, all that MHC needs to do to share benefits is to make the test available to members of the population at a nominal fee. Thus, in many cases the best form of compensation to a community or population will be to make the test, treatment or other application reasonably available to members of the population or community. In other cases, however, companies may need to offer individual subjects additional compensation. How much compensation is owed would be a function of the total benefits created from the research and development and the contributions of the various parties. Individual subjects have the best case for demanding financial compensation when 1) the profits are high, 2) individual subjects have made substantial contributions to the research. The Moore case would meet these two criteria. The Canavan case, on the other hand, might not, since the profits will probably not be very high and no individual subject made a substantial contribution to the research; the community as a whole made the contribution.

To summarize these two cases, researchers and research sponsors have substantive duties as well as procedural duties relating to benefit sharing in genomics. Principles of substantive justice require that researchers share benefits according to the contribution of a person of population: the greater the contribution they make to the research, the greater share of benefits they deserve. Principles of procedural justice support the idea that researchers should develop specific plans for benefit sharing and they should discuss those plans with the subjects and populations.

Let's move beyond these local cases and consider a global benefit sharing related to the commercialization of the human genome. From a utilitarian (or cost–benefit) perspective, the commercialization of the genome is reasonable and justifiable, since the probable benefits of commercialization for science, technology and society outweigh the probable harms (Resnik 2003). Nevertheless, one should still ask questions about the overall pattern of the distribution of the benefits and burdens resulting from the commercialization of the genome. What is a fair or just way to share the benefits of commercialization among people within a nation and among the people of the world? How should we address the concern that the commercialization of the genome will increase the socio-economic gap among developed nations and developing nations as well as the gap between rich and poor people within nations? What should we do to ensure access to genetic information and technology?

These are highly complex questions that touch on a variety of practical issues, such as insurance, discrimination, privacy and testing, and involve in-depth inquiries into to different theories of justices, such as egalitarianism, libertarianism and utilitarianism (Mehlman and Botkin 1998; Buchanan et al. 2000). I cannot hope to answer all of these difficult questions here. However, I would like to address an issue relating to the commercialization of the human genome: will commercialization increase the gap between rich and poor? A number of different commentators have expressed the concern that the commercialization of the human genome will increase the gap between the rich and the poor (Andrews and Nelkin 2001; Cahil 2001). They are concerned that the benefits of commercialization are flowing directly toward private companies and researchers in the developed world but not to the developing world. This critique of the commercialization of genetic research is not entirely new and expresses the same kinds of concerns that people have had about a variety of new technologies, including personal computers, television and automobiles. In each of these cases people worried that only the rich people would be able to afford the new technologies and, therefore, the benefits would not be shared fairly because they would accrue to the rich and not the poor.

To gain some insight into the fairness (or unfairness) of the gap between rich and poor let us consider a theory of justice developed by the late John Rawls. Rawls' theory has had a huge influence on social and political philosophy in the last three decades, and many people have applied his insights to the distribution of health and healthcare (see, for instance, Daniels 1985). Rawls' theory is known as a social-contract theory, because it holds that principles of justice are the rules for governing society that hypothetical parties would accept, provided that they are placed behind a veil of ignorance that prevents them from knowing who they are in the society they are forming. The rules adopted by these hypothetical parties would be like a contract for forming a just society. According to Rawls, the contractors would adopt two basic rules: 1) fundamental moral and legal rights should be distributed equally, and 2) socioeconomic goods may be distributed unequally provided that (a) the unequal distribution makes everyone in society better-off, especially the worst-off members, and (b) there is fair equality of opportunity in society (Rawls 1971). The first principle is known as the equality principle; the second is known as the difference principle. If we apply Rawls' principles to the commercialization of the genome, we should ask the following question: will the commercialization of the human genome make everyone in society better-off without violating moral or legal rights? If the answer is 'yes' to this question, then commercialization is just.

Critics of DNA patenting argue that commercialization is unjust because patenting increases the price of genetically based tests and treatments by giving the patent holder a limited monopoly on his product or process (Andrews and Nelkin 2001). Unless competitors can develop inventions that 'work around' the patent, they will not be able to enter the market until the patent expires, and the cost of product or process will remain high until the patent expires. Most of these profits will benefit the large corporations that own DNA patents. Critics of DNA patenting point to the high costs of genetic tests, such as Myriad's BRCA1 test, and the high cost of genetic medicines, such as clotting factors or erythropoietin, as evidence of the injustice of patenting. Some critics argue that the best way to increase access to the genome and promote justice is to ban patents on all DNA (Rifkin 1998).

This argument makes several mistakes and oversights, however. The argument ignores the fact that the patent period lasts only 20 years, half of which usually occur when a product or service is undergoing clinical testing. For example, in the pharmaceutical industry it usually takes about 10 years and $500 million to develop and test a new drug and bring it to the market (Goldhammer 2001). This means that a company has about 10 years to earn back its investment. During this time, a company will charge what the market will bear, because it knows that its profits will diminish greatly once the patent expires. Once the patent on a drug expires, other companies can make generic versions of the drug, at a great savings to consumers. Costs will continue to fall as a result of improvements in manufacturing and economies of scale. If the company did not expect that it would have patent protection, it would not have invested its money in developing the drug, and the drug may have never entered the market. In the short run, patenting interferes with access to medications, but in the long run it increases access to medications by providing inventors and investors with incentives to conduct and sponsor research. Since the patent system grants inventors a temporary monopoly, it tends to produce short-term problems with access to technology, but its long-term effects promote access by stimulating investment in research and development.

The history of science and technology contains examples of many products and services that were initially very expensive – and therefore available only to the rich – that soon fell in price. Automobiles, refrigerators, microwave ovens and personal computers at one time were so expensive that they were available only to rich people in developed countries. Today, almost everyone in a developed country has access to these products, and many people in developed nations have access to the products. The point here is that new technologies can create a temporary gap between rich and poor, but that gap narrows over time. If the history of science and technology offers us any useful lessons for the DNA-patenting debate it is that the commercialization of the human genome will probably promote global benefit sharing in the long run, because it will encourage investment in genetic technologies that will eventually be widely available.

Opponents of DNA patenting may argue that the success of the patent system is overrated. Patents do not always lead to long-term benefits for society and may do more harm than good. Researchers and companies can abuse the patent system by using patents to block downstream research, by refusing to grant license, and by attempting to extend the life of their patents by 'double patenting' or other illegal activities. This is an empirical debate that cannot be resolved here. Economists and legal scholars continue to debate the social utility of the patent system; however, there is a general consensus that it plays a key role in promoting the development of science and technology, which benefits society. Thus, the patent protections that create problems with access to technology can be justified on the basis that they produce good consequence in the long run.

However, there are some exceptions to this patent-protection policy. In some cases the short-term inequities may be so unjust that countries are justified in restricting patent rights in order to make products or services readily available. For example, the HIV/AIDS epidemic in sub-Saharan Africa is a public-health crisis of such grave proportions that countries are morally justified in restricting or overriding patents on essential HIV/AIDS medications in order to increase access to these medications by lowering their cost (Resnik and De Ville 2002). In some rare instances the need to address inequities is so great that governments can set aside the laws that normally govern patenting. However, governments should use great discretion and care in applying this emergency exception to patents to avoid treating every problem as a crisis.

Chapter 13

Conclusion

This essay has argued that the human genome is not literally our common heritage. If the human genome were literally our common heritage, the patenting of human DNA would be morally unacceptable because it would require the consent of every human being, a practical impossibility. Even though the human genome is not literally our common heritage, it is still a very important common resource, and we have moral duties of stewardship and justice vis-à-vis the human genome. Our duties of stewardship include duties to refrain from harming the human genome but not duties to benefit the genome actively, because the idea of 'benefiting' or 'improving' the genome has clear eugenics implications. Our duties of justice imply obligations to share benefits fairly in genetics research and development. Benefit sharing can take place at a local level when researchers develop treatments or tests that become reasonably available to the populations or communities that participate in research. Local benefit-sharing obligations require researchers to provide financial compensation to participants only in rare instances where researchers and companies stand to profit a great deal from the tissues collected from a single person or small group of people. Local benefit-sharing obligations also require researchers to develop plans for sharing benefits and for discussing these plans with study populations. Finally, global benefit sharing may occur as products and services developed by companies become less expensive and more widely available. Short-term problems with access to genetic technology can be justified on the grounds that the system that allows such inequities, i.e. the patent system, promotes the interests of all members of society, especially the worst-off members, in the long run.

References

The American Heritage® dictionary of the English language, 2000. 4th edn. Houghton Mifflin, Boston. [http://www.bartleby.com/61/]
Andrews, L. and Nelkin, D., 2001. *Body bazaar: the market for human tissue in the biotechnology age*. Crown Publishers, New York.
Buchanan, A., Brock, D., Daniels, N., et al., 2000. *From change to choice: genetics and justice*. Cambridge University Press, Cambridge.
Cahil, L.S., 2001. Genetics, commodification, and social justice in the globalization era. *Kennedy Institute of Ethics Journal*, 11 (3), 221-238.
Council of Europe, 2001. *Recommendation 1512: Protection of the human genome*. [http://assembly.coe.int/Documents/AdoptedText/ta01/EREC1512.htm]
CRG, 2000. *The genetic bill of rights*. Council for Responsible Genetics CRG, Cambridge. [http://www.gene-watch.org/programs/bill-of-rights/bill-of-rights-text.html]
Daniels, N., 1985. *Just health care*. Cambridge University Press, Cambridge.
Diamond vs. Chakrabarty, 1980. 447 U.S., 303 (1980).
Doll, J.J., 1998. The patenting of DNA. *Science,* 280 (5364), 689-690.
Eisenberg, R.S., 1990. Patenting the human genome. *Emory Law Journal,* 39 (3), 721-745.
Foubister, V., 2000. Gene patents raise concerns for researchers, clinicians. *American Medical News,* 43 (7), 21 Feb. [http://www.ama-assn.org/amednews/2000/02/21/prsb0221.htm]
Goldhammer, A., 2001. Current issues in clinical research and the development of pharmaceuticals. *Accountability in Research,* 8 (4), 283-291.

Greenberg v. Miami Children's Hospital Research Institute, Inc., 2003. WL 2122463472003 (S.D. Fla. 2003).

Gutmann, A. and Thompson, D., 1996. *Democracy and disagreement*. Belknap Press, Cambridge.

Human Genome Organization Ethics Committee, 2000. Genetic benefit sharing. *Science,* 290 (5489), 49.

International Federation of Gynecology and Obstetrics, 1997. *Patenting human genes.* Available: [http://www.figo.org/] (15 March 2004).

Juengst, E.T., 1998. Should we treat the human germ-line as a global human resource? *In:* Agius, E. and Busuttil, S. eds. *Germ-line intervention and our responsibilities to future generations.* Kluwer Academic Press, Dordrecht, 85-102.

Karp, J.P., 1991. Experimental use as patent infringement: the impropriety of a broad exception. *Yale Law Journal,* 100 (7), 2169-2188.

Knoppers, B.M., 2000. Population genetics and benefit sharing. *Community Genetics,* 3 (4), 212-24.

Kolata, G., 2000. Who owns your genes? *The New York Times* (15 May 2000).

Looney, B., 1994. Should genes be patented? The gene patenting controversy: legal, ethical, and policy foundations of an international agreement. *Law and Policy in International Business,* 26, 231-272.

Mehlman, M.J. and Botkin, J.R., 1998. *Access to the genome: the challenge to equality.* Georgetown University Press, Washington DC.

Miller, A.R. and Davis, M.H., 2000. *Intellectual property: patents, trademarks, and copyright in a nutshell.* West Group, St. Paul.

Moore vs. Regents of the University of California, 1990. 793 P.2d 479 (Cal. 1990).

Nuffield Council on Bioethics, 2002. *The ethics of patenting DNA: a discussion paper.* Nuffield Council on Bioethics, London.
[http://www.nuffieldbioethics.org/filelibrary/pdf/theethicsofpatentingdna.pdf]

Ossario, P., 1998. Common heritage arguments and the patenting of DNA. *In:* Chapman, A.R. ed. *Perspectives on gene patenting.* American Association for the Advancement of Science, Washington DC, 89-110.

Patent Act 35 USC 101, 1952, 1995.

Rawls, J., 1971. *A theory of justice.* Belknap Press, Cambridge.

Rawls, J., 1993. *Political liberalism.* Columbia University Press, New York.

Resnik, D.B., 2001a. DNA patents and human dignity. *The Journal of Law, Medicine, and Ethics,* 29 (2), 152-165.

Resnik, D.B., 2001b. DNA patents and scientific discovery and innovation: assessing benefits and risks. *Science and Engineering Ethics,* 7 (1), 29-62.

Resnik, D.B., 2002. Discoveries, inventions, and gene patents. *In:* Magnus, D., Caplan, A. and McGee, G. eds. *Who owns life?* Prometheus Books, Amherst, 135-160.

Resnik, D.B., 2003. *Owning the genome: a moral analysis of DNA patenting.* State University of New York Press, Albany.

Resnik, D.B. and De Ville, K.A., 2002. Bioterrorism and patent rights: compulsory licensure and the case of Cipro. *American Journal of Bioethics,* 2 (3), 29-39.

Rifkin, J., 1998. *The biotech century: harnessing the gene and remaking the world.* Tarcher/Putnam, New York.

Rolston, H., 1994. *Conserving natural value.* Columbia University Press, New York.

Spectar, J.M., 2001. The fruit of the human genome tree: cautionary tales about technology, investment, and the heritage of mankind. *Loyola of Los Angeles International and Comparative Law Review,* 23 (1), 1-40.

Sturges, M.L., 1997. Who should hold property rights to the human genome? An application of the common heritage of mankind. *American University International Law Review,* 13 (1), 219-261.

UNESCO, 1997. *Universal declaration on the human genome and human rights.* United Nations Educational, Scientific, and Cultural Organization UNESCO, Geneva. [http://www.unesco.org/shs/human_rights/hrbc.htm]

United States Constitution. Article 1, section 8, clause 8 (1787). [http://www.constitution.org/indexco_.htm]

Venter, J.C., Adams, M.D., Myers, E.W., et al., 2001. The sequence of the human genome. *Science,* 291 (5507), 1304-1351.

CONCLUSIONS

14
Conclusions: Towards ethically sound life sciences

Michiel Korthals[#]

Introduction

Life sciences concentrate on life and death: this simple statement stands for most of the urgent ethical problems these sciences are confronted with. Apparently there is a strong connection between studying living organisms, the way they reproduce, age and die, and social and ethical considerations triggered by them. In whatever way their research results are published in scientific journals, however they propose to explain the theories on how plants, animals or humans become ill and can be protected against diseases, whatever their advise on crop improvement or health mechanisms, the life sciences cannot escape from ethical issues, controversies, dilemmas even, that require debate, consultation, guidelines, intuition, experiences and so on.

In this collection of papers we have intensively discussed the new, and often uncertain impacts of these sciences and their connected technologies, as well their wider (global) impact. It has become clear that many ethical issues are not only triggered by possible misconduct in the treatment of animals or humans, but also by these uncertain and wider impacts. So, the ethical chapters do not concentrate on wrongdoings of scientists at all, but on the identification of possible harm and disturbances in society that at least are influenced, and in many cases even directly determined, by these sciences. For example, genetic information can play a part in disturbing the traditional privacy regulations or in challenging patients and consumers to take their future into account (e.g., in the case of information on vulnerability to cardiovascular diseases).

After the systematic and concentrated effort on separate issues, it is now time to look at it from a more distant point of view. First, let us look at the tasks of ethics of the life sciences, secondly, at its role in the broader field of ethics and in that of philosophy in general, and thirdly, at its role in society. I will take this step of reflection by first summarizing the main issues of this collection of studies, then sketch some wider relations of ethics of the life sciences and pose some future questions and challenges.

Ethical issues of conducting research (internal to science)

We started this collection with issues internally related to conducting science, like the ethical role of scientists (professionals) in organizations. Subsequently, we tackled issues of research and the publication of research. Many of these issues circle around problems of integrity and how to prevent misconduct in conducting science. However,

---

[#] Applied Philosophy Group, Department of Social Sciences, Wageningen University, Hollandseweg 1, 6706 KN Wageningen, The Netherlands. E-mail: Michiel.Korthals@wur.nl

the management and access of scientific data have changed by, firstly, the new technologies of both information acquiring and information organization (digitalization of science), and secondly, the new private–public co-operations in the field of research.

Animal welfare and the treatment of human subjects, collegiality (not sexist, racist etc.) were all discussed. The need of attention for misconduct highlights the important ethical role of scientific integrity and corresponding types of responsibility as they are developed in the course of scientific practices (Harris, Pritchard and Rabins 2000; Maker 1994). So, good practices provide guidelines for scientists in cases of dilemmas and new moral challenges. On the side of the human subjects, either as subject to experiments or as subject to large-scale epidemiological observations, informed consent is often an important ethical regulatory device. However, the private–public co-operation very often leads to commercialization of academic interests, which means that more and more scientists are confronted with conflicts of interest. This becomes clear when questions arise on who owns the intellectual property (patents) and who decides the research agenda (research priorities). Also conflicts of commitment, like diminishing attention for teaching or for the public role of the scientists, can become disturbing complications of these new relationships.

All these new developments indicate that societal relationships with the life sciences are very dynamic and produce new controversies, dilemmas and ethical problems. This implies that indeed existing role models and best-practice guidelines are not always sufficient to give a transparent, honest and responsible answer. Most of the existing role models and best-practice guidelines concentrate on 'informed consent'. However, because ethical issues of misconduct and catastrophes are surpassed by new, urgent ethical issues like that of embedding life sciences and technologies in socio-cultural contexts, and because of the restricted scope of 'informed consent', we have to look for other ethical perspectives. The need of attention for misconduct and for scientific integrity can, in this stage of the life sciences, no longer be answered by taking into account the (existing) role of responsibility of life sciences. Public consultations and intense communication with stakeholders can give some answers to these disturbing problems, whilst other ways of tackling ethical issues are yet to be experimented with. However, the challenge is to reconstruct the role of the life-science professional who will have to take these developments into account.

Science and societal problems

Besides the attention for misconduct of scientists, scientific catastrophes and disasters, ethics for life sciences should concentrate on the identification, analysis and evaluation of the most important uncertainties and disturbances that these sciences and technologies confront us with. It is a truism that modern life sciences, besides covering knowledge of the natural world, also cover knowledge of the social world, be it in naïve or sophisticated form, and do not only transform nature, but also maintain and amend existing systems of natural and social transformation. Conducting research on seeds, cows, harvesting machineries or genes, includes working on social organizations.

Life sciences and technologies are not just an instrument or a nice toy for researchers, because they change humans and their relationships as well. These changes concern both social and cultural aspects. The cultural influence can be seen in the impact of these sciences on the way people perceive their lives and bodies; e.g. because of the increasing attention to genomics, people are becoming more and more aware of their genetic make-up affecting their behaviour and functions. The social aspects concern, for example, challenges to traditional versions of social responsibility of the life sciences and what their scope should be and how the relationship should be between individual, social and public responsibility. New forms of public participation and consultation are enacted, and the challenges are, *inter alia*, what the scheme of reference, the framework and the procedure will be, in order to be able to take into account all conflicting interests. Life scientists are no longer exempt from participating in public debates and no longer allowed to carry out research on their own. They should engage themselves in enhancing the quality of the public debate and learn from participation in public debates. This comprises what is often called co-evolution of science and society: all groups involved, even with conflicting interests and opinions, can try to develop new ways of understanding and interacting.

The most urgent problems in this regard concern the increasing disagreement between the food and agricultural professions and large parts of society on the way medicine, food and agriculture should be viewed, assessed and organized. This disagreement is partly internalized by these professions, which means that the professions are split up in rivalling parties that participate in sometimes radically different technological systems. However, for the professions in general it means, firstly, that they are confronted with a *lack of trust* from the side of their end users, society. Secondly, the food and agricultural science and engineering profession are confronted with a *split between two or more technological systems* that are partly at odds with each other and with which different types of science and engineering are connected. The challenge of the professions is to manage the peaceful coexistence of these two competing technological systems.

Let me illustrate these two radical theses. Although the recent criticisms started in the nineteen sixties with the alarming messages on the environment in Rachel Carson's book Silent Spring, it was the biotechnology wave, from the early seventies, that really sparked a booming wave of discontent with the general goals, standards, skills, research priorities and values of these high-tech professions. Damage to the environment in the form of pollution, or decrease in (agro-)biodiversity, deterioration of the aesthetic quality of the landscape and food, increasing animal-welfare problems (like broiler chickens or factory pigs) were attributed to a large-scale intensive technological system, in which agricultural and food scientists and technologists played a crucial role. Because of the gap between producers and consumers with respect to food production, it took some time before consumers realized what was going on in food production. Since the nineties, after several food crises, mass media paid attention to these circumstances, and made many consumers conscious of the material and immaterial costs of this large-scale intensive farming system. They questioned the aims, standards and competencies of a profession that contributed to these developments.

These criticisms became more or less standard in the nineties during frequent crises in the food-production chain, like swine fever, foot-and-mouth disease, BSE crisis, avian influenza and the dioxin scandal. The belief of consumers in the reliability of the people directly involved in the food chain, whether they were industrialists, government agencies or food professionals working for food industries, sunk to a record low. On the other hand, the belief in non-food organizations like Greenpeace has increased. Integrity, transparency, elsification (paying attention to ethical, legal and social aspects [ELS] of science and technology), PPP (planet, people, profit), CRS (corporate social responsibility) are some of the reactions of business and science circles. Large scientific organizations, like Wageningen University and its alumni organization of agricultural and food professionals, are trying to rewrite their codes, and to pay attention to professional ethics. In teaching and education, agricultural and food engineers learn competencies to tackle societal and ethical dilemmas. All these developments make one thing clear: the values, standards and competencies of the food and agricultural professions are not sufficiently endorsed by society at large. What should agriculture, food, plant and animal technologists, engineers and scientists do? The relationship between these professions and society is under severe strain.

In reaction to these critical developments, many engineers took the criticisms seriously, and started to think in new ways. Small-scale agricultures, with tailor-made technologies in eliminating pests and viruses and enhancing yields with local participatory organizations were institutionalized. Both in the developing and in the developed world artisan ways of farming and food production (i.e. organic agriculture or slow food!) are transformed in this extensive way. In many developed countries these systems are a small minority, but in others still the dominant form.

These developments bring about my second remark, on the coexistence of at least two different agricultural and food systems confronting the professions. This coexistence is not always a peaceful one. Be it GM agriculture versus non-GM agriculture (and food) or intensive versus extensive farming systems, these systems have different types of crop protection and soil protection, different types of zoonoses (veterinary diseases that can also affects humans) and different types of what can be called 'contamination'. Non-GM crops can be contaminated by GM seeds; organic farming feels threatened by pesticides of non-organic farming, and non-organic farming is contaminated by organic technologies (there was a case of *Phytophthora* in The Netherlands in 2003). The use of *Bt* genes in maize or potato makes it more difficult for both systems to stay ahead in the race against bugs, because resistance is built up more rapidly. The struggle between ocean fishing and aquaculture (farm fishing) is another example of the difficult coexistence of different systems. These systems use all kinds of rivalling technologies, which exist simultaneously but often cause great social and natural conflicts. The EU has enacted special regulation on coexistence, but nobody understands exactly how to tackle these issues. At the end of the food chain we have similar subsystems connected to two different types of consumers: the obese with the fast-food and biotech sectors, the more conscious consumer with the others. Anyhow, the existence of two or even more agricultural and food systems is one of the most intriguing challenges for the professions, and it makes value conflicts the core issues of their undertakings (Charles 2001; Pollan 2001; Maker 1994). I hope that life scientists and engineers can contribute in making the warlike aspects of this coexistence into peaceful aspects and will learn from this controversy.

Because of the dynamics of technologies and the dynamics of societies in which these technologies are realized, we should look for new kinds of interactions between the agriculture and food professions and society. The relationship between the two will become more and more unclear and we need ethicists, but not only them, to map the grey zones and potential learning processes. It is not possible to fix, once and for all, what the aims and quality standards of a practice or a profession are. But we should look for new fluent and robust ways of rationally dealing with ethical-technological problems.

Both Professional Ethics and Applied Ethics have something to deliver in tackling these problems. So, let us look at elements we can use, like utilitarian and deontological reasoning patterns, value dilemmas, future scenarios with varying value–technology dimensions, ethical stakeholder analysis, public consultations and other deliberative methods of empirical philosophy. Core question should be how life-science and engineering practices should respond to societal values, and how technologies can contribute positively to values of the living world.

The status of ethics of life sciences

If ethics of life sciences can make a difference and has such an important task in contributing to the improvement of the interface between the life sciences and society, it is also necessary to reflect upon its methods, goals and status in general. To what extent does ethics for life sciences have an independent status, what can it learn from academic ethics or can this applied-ethics branch even act as a role model for the more formal and principled branch of ethics? What kind of impact can applied ethics have on academic ethics and what can academic, general ethicist learn from ethics of life sciences? More broadly viewed, what kind of social role should ethics for life science have?

Traditionally, applied ethics is seen as applying given principles from the mother discipline, fundamental ethics, to a concrete problem; some empirical facts are considered necessary to get the applicable conclusion. However, as early as the nineteen fifties, it became clear to many ethicists working on the field of medicine, that their applied work was different and that in applying fundamental norms, new insights with respect to these fundamental norms where acquired. Also new fields of application emerged that were not foreseen when the fundamental norms were formulated and justified. It turned out that many applied ethics were not applying existing norms but constructing new norms, whether as in English common law, by progressing from precedents, or more boldly, by proposing whole new norms and procedures. As a matter of fact, the first category was often confronted with the reproach that it only tried to adapt public opinion to the whims of scientists and technologists (or industrialists). However, it became more and more clear that ethics of life sciences could indeed point at new insights and was not only a kind of social ambulance service by producing order out of chaos and removing the rubbish for the triumphant march of the life sciences. It turned out that ethics of life sciences has some influence on societal views on the life sciences and on enhancing the ethical debate on the life sciences. It is clear that society is becoming more and more aware of the fact that the contemporary discussion on genomics has nothing to do with the discussion on eugenics sixty years ago.

The new insights of applied ethics that should be taken seriously by fundamental ethics are not connected with the committees and endless procedures that often are inaugurated without any sign of really improving the ethical debate, but in the alternatives for informed consent, like new types of public consultations, in debunking traditional distinctions of fundamental ethics like that between citizen and consumers (Korthals 2004), and in making clear the importance of our biological heritage and its role for future generations. Although recent acquirements are important, the main work is still to be done, and herein lies the most challenging work of applied ethics, in particular in the field of sound frameworks for debates with conflicting, often incommensurable interests and visions.

Chapter 14

Conclusion

In the last decades the ethical issues within the life sciences have been widely discussed in journals and textbooks. Professional integrity, accountability and responsibility are outlined by prominent ethicists as important features of ethically sound life sciences, at least with respect to issues within science. However, the ethical issues that are connected with the interface of life sciences and society are not discussed to such an extent, often not even identified at all. The priorities of research, the impact of various life-science alternatives on society and the restoration of trust in the life sciences are usually neglected. If they are discussed, more often than not, a kind of societal consensus vis-à-vis the interface of life sciences and society is assumed. However, consensus and harmony cannot be assumed; we must begin with deeply conflicting, often incommensurable opinions on the future of this relationship.

How to start co-operation, taking into account this difference in interest and opinion? What should be the main issues? Who should be the main parties and stakeholders? Does everyone have a voice? How to deal with accountability and responsibility of the life sciences in this respect? What capacities should life scientists have in participating on the edge of this interface? How should the curriculum be modified to meet these societal demands? What criteria can help the consumer/citizen in his decision to trust new activities of the life sciences? What can ethics do to contribute to these trust enhancing measures? These are all questions that need to be answered.

Recommended literature

Charles, D., 2001. *Lords of the harvest: biotech, big money, and the future of food*. Perseus Publishing, Cambridge.
Harris, C.E., Pritchard, M.S. and Rabins, M.J., 2000. *Engineering ethics: concepts and cases*. 2nd edn. Wadsworth, Belmont.
Hayden, C., 2003. *When nature goes public: the making and unmaking of bioprospecting in Mexico*. Princeton University Press, Princeton.
Keulartz, J., Schermer, M., Korthals, M., et al., 2004. Ethics in technological culture: a programmatic proposal for a pragmatist approach. *Science, Technology and Human Values*, 29 (1), 3-29.
Korthals, M., 2004. Ethics of differences in risk perceptions and views on food safety. *Food Protection Trends*, 24 (7), 30-35.
Maker, W., 1994. Scientific autonomy, scientific responsibility. *In:* Wueste, D.E. ed. *Professional ethics and social responsibility*. Rowman and Littlefield Publishers, Lanham.
Pollan, M., 2001. *The botany of desire: a plant's-eye view of the world*. Random House, New York.

List of authors

M. Boon	Associate Professor, Department of Philosophy, University of Twente, Enschede, The Netherlands
R.E. Bulger	Professor, Department of Anatomy, Physiology and Genetics, Uniformed Services University of the Health Sciences, Bethesda, MD, USA
R. Chadwick	Professor, Furness College, Lancaster University, Lancaster, United Kingdom
T. de Cock Buning	Professor, Department of Biology and Society, Free University, Amsterdam, The Netherlands
M. Düwell	Professor, Department of Philosophy, University of Utrecht, Utrecht, The Netherlands
B. Gert	Professor, Department of Philosophy, Dartmouth College, Hanover, NH, USA
M. Häyry	Professor and Head of Centre for Professional Ethics, University of Central Lancashire, United Kingdom
R. Heeger	Professor, Department of Theology, University of Utrecht, The Netherlands
R. Jeurissen	Professor, EIBE – Institute for Responsible Business, Nyenrode University, Breukelen, The Netherlands
J. Keulartz	Associate Professor, Applied Philosophy Group, Wageningen University, The Netherlands
J.H. Koeman	Emeritus Professor of Toxicology, Wageningen University, The Netherlands
M. Korthals	Professor, Applied Philosophy Group, Wageningen University, The Netherlands
H. Letiche	Professor, University for Humanist Studies, Utrecht, The Netherlands
D.B. Resnik	Professor, Department of Medical Humanities, The Brody School of Medicine, East Carolina University, Greenville, NC, USA
P.B. Thompson	Distinguished Professor, Department of Philosophy, Michigan State University, East Lansing, MI, USA
M.A.J.S. van Boekel	Professor and Chair of Product Design and Quality Management Group, Department of Agrotechnology and Food Sciences, Wageningen University, The Netherlands
H. van den Belt	Assistant Professor, Applied Philosophy Group, Wageningen University, The Netherlands
A.J. van der Zijpp	Professor, Animal Production Systems Group, Department of Animal Sciences, Wageningen University, The Netherlands
J. Wempe	Director, CSR Resource Centre, and Professor, Department of Business Management, Rotterdam School of Management, Erasmus University, Rotterdam, The Netherlands
H. Zandvoort	Associate Professor, Department of Philosophy, Delft University of Technology, Delft, The Netherlands
H. Zwart	Professor, Department of Philosophy and Science Studies, Faculty of Science, Mathematics and Computer Science, University of Nijmegen, The Netherlands

Printed in the United States
38785LVS00002BA/91-102